Cultural Studies of Science Education

Volume 15

The series is unique in focusing on the publication of scholarly works that employ social and cultural perspectives as foundations for research and other scholarly activities in the three fields implied in its title: science education, education, and social studies of science.

The aim of the series is to establish bridges to related fields, such as those concerned with the social studies of science, public understanding of science, science/technology and human values, or science and literacy. Cultural Studies of Science Education, the book series explicitly aims at establishing such bridges and at building new communities at the interface of currently distinct discourses. In this way, the current almost exclusive focus on science education on school learning would be expanded becoming instead a focus on science education as a cultural, cross-age, cross-class, and cross-disciplinary phenomenon.

The book series is conceived as a parallel to the journal Cultural Studies of Science Education, opening up avenues for publishing works that do not fit into the limited amount of space and topics that can be covered within the same text. Book proposals for this series may be submitted to the Publishing Editor: Bernadette Ohmer E-mail: Bernadette.Ohmer@springer.com.

More information about this series at http://www.springer.com/series/8286

Kathrin Otrel-Cass • Martin Krabbe Sillasen
Auli Arvola Orlander
Editors

Cultural, Social, and Political Perspectives in Science Education

A Nordic View

Springer

Editors
Kathrin Otrel-Cass
Department of Learning and Philosophy
Aalborg University
Aalborg, Denmark

Martin Krabbe Sillasen
VIA University College
Teacher Education
Silkeborg, Denmark

Auli Arvola Orlander
Stockholm University
Stockholm, Sweden

ISSN 1879-7229 ISSN 1879-7237 (electronic)
Cultural Studies of Science Education
ISBN 978-3-319-61190-7 ISBN 978-3-319-61191-4 (eBook)
DOI 10.1007/978-3-319-61191-4

Library of Congress Control Number: 2017954383

Printed on acid-free paper

This Springer imprint is published by Springer Nature
The registered company is Springer International Publishing AG
The registered company address is: Gewerbestrasse 11, 6330 Cham, Switzerland

Contents

Chapter 1
Introduction

Auli Arvola Orlander, Martin Krabbe Sillasen, and Kathrin Otrel-Cass

Why did we write this book and why should it be of interest to our readers? 'We' refers to a community of scholars coming from the Nordic countries, including Denmark, Sweden and Norway. Our readers, those who we anticipate, are scholars like us with their work relating to science education but also those in teacher education and of course anyone who feels attracted and hopefully inspired by our vexations. The community of writers who contributed to this work are diverse, intersecting and questioning science education teaching, learning and teacher professional development from different angles and with different agendas. Amongst our community were a number of problematic issues regarding scientific inquiry, socio-materialism, the science lab, private enterprises in schools, outcome of teacher professional development, citizenship, norms and values that we agreed on, and that became the driving force for writing this book.

With this book we want to problematise selected fields of research in science education and teacher professional development. As a community we felt that much time and effort is still devoted to seek out 'best practice' to address problems in science education. We also wondered about the continued research and policy focus on the individual student to being successful and finding school science meaningful and interesting. For example, research focus is on topics including argumentation,

A.A. Orlander (✉)
Stockholm University, Stockholm, Sweden
e-mail: auli.arvola.orlander@mnd.su.se

M.K. Sillasen
VIA University College, Teacher Education, Silkeborg, Denmark
e-mail: msil@via.dk

K. Otrel-Cass
Department of Learning and Philosophy, Aalborg University, Aalborg, Denmark
e-mail: cass@learning.aau.dk

K. Otrel-Cass et al. (eds.), *Cultural, Social, and Political Perspectives in Science Education*, Cultural Studies of Science Education 15,
DOI 10.1007/978-3-319-61191-4_1

inquiry-based learning and scientific modelling, context of students' learning, science teaching and students' conceptual learning (Lin et al. 2014).

We take a particular interest in the factors that go beyond the individual and shape the conditions for science education in late-modern classrooms. Another important aspect is our shared concern that the story *about* science in science education may still be that science is value free and neutral since it is unclear when the opposite is being told. We wonder whether the reasons for this have been that understanding the conditions for teaching and learning in science have focused too narrowly on the interactions between teachers and students (Alton-Lee 2003). Instead we believe that it is necessary to widen the focus including to investigate how cultural norms or political ideologies shape science education and science teachers' professional development. This interest is spawned also in the our community of researchers, who come mainly from Nordic countries that are at times portrayed as exemplary in embracing high standards of democratic and social values. We wondered whether this 'Nordic' ideology may create the illusion of latent justice in science education and that students discovering science for themselves was simply a question of their motivation and curiosity, depending on them and their teachers to make it happen, not including the idea that science education is value-laden.

The chapters collected in this book reflect the emergence and continuation of critical thinking in science education organised by themes of cultural, social and political perspectives in science education. Selected chapters present new takes on theoretical frameworks and units of analysis that take into account that science education is socially, culturally and politically shaped. From macro to micro perspectives, we review the role of the structures that bind or separate people, objects and ideas in science education and also highlight that science education, including its conceptual and material constituents, is neither innocent nor neutral.

1.1 Our Intentions

We made it our challenge to 'trouble' existing frameworks and conditions for researchers, teachers and learners in science classrooms, which include political, societal and cultural issues, professional development frameworks and the nature of educational materials. The authors raise questions about the conditions for identity formation, participation and learning in science. Some trouble the process of science education research through the use of different theoretical framework, others are concerned with current understandings of scientific inquiry, the challenges of involving private companies or the place for human-centred perspectives in science education. The positive outcome of teacher professional development programmes is questioned for excluding different teacher voices and experiences in the quest to reinforce its own success.

The troubling issues presented in various ways in the different chapters are related to how young people and teachers react to the social-cultural-political framing of science education. When 'Science for All' continues to be the ideal, that is, being promoted, it perpetuates educational models that link economic success of countries with the number of individual students who master international tests like PISA but continue to produce statistics of failure (e.g. Serder and Ideland 2016). This raises the question of who will be excluded by such good intentions that create images of desirable students with desired behaviour.

1.2 Emergence of a Community of Critical Reflective Researchers

The idea for this book grew out of a network that was funded in the first phase by the Swedish National Bank and in the second phase by the Crafoord Foundation. We are thankful for their financial support that enabled authors and invited guests to meet and discuss various issues regarding science education. What we were seeking was to build a research community that developed a consciousness and autonomous thinking, reflective and self-critical abilities and empathetic responses as the ultimate purpose for all science education research towards democratic citizenship.

To inspire the network's critical thinking about science education, we invited interesting researchers who shared and discussed their results with us. Those discussions troubled different assumptions about the prevalent content/discourse in science education and nourished the thinking and wonderings we had.

Elizabeth de Freitas, from Manchester Metropolitan University (UK), gave us new insights into new materialist ontologies in education. She was also a lecturer in a PhD course that *Paola Valero*, Stockholm University (Sweden), arranged about bodies in science education that was held in connection to one of the network meetings held in Aalborg, Denmark. A course that many of the network's PhD students and researchers participated in. At the same occasion, *Marie Öhman*, from Örebro University (Sweden), shared her work with the network about exercising bodies in education, relating also to questions of power and science education. This occasion created a space and opportunity for the authors of this book to trouble and think about the notion of body in science, a theme that was picked up on a number of occasions in the book.

Sharon Todd from Maynooth University in Ireland visited the network in Stockholm (Sweden) and gave a lecture entitled 'Creating Pedagogical Spaces of Transformation: The Difficult Task of Facing Humanity'. She discussed issues concerning political and ethical aspects of education, issues of democracy and cosmopolitanism and images of femininity and masculinity in educational knowledge. In the same meeting, *David Kronlid*, from Uppsala University, Sweden, presented his reflections on 'Transformative Learning' with focus on environmental ethical research, mobility and climate equity and education in sustainable development. Their work inspired the continued discussions on the political dimensions to science and in particular environmental education, as not so innocent discourses.

At a meeting in Malmö (Sweden), *Heidi Carlone* from the University of North Carolina at Greensboro (USA) discussed her work and content in the article 'The cultural production of science in reform-based physics: Girls' access, participation, and resistance'. Inspired discussions and perpetuations on the topics of gender in science followed.

These discussions but also the ongoing dialogue we had with each other were instrumental to the process of thinking to the contributors of this book. The content of the book can be regarded as a reification of the ongoing process.

1.3 Organisation of the Book

After this introduction we will present a reflection on the conceptualisation process for this book, by presenting an article in cartoon format. The intention here is also to trouble traditional ways of presenting and sharing reflections to the research community, to highlight that a visual argument embeds additional and different meanings to written arguments. On reflection and due to the community we become, we found it very difficult to organise our thinking into domains since they are entangled entities. We find it therefore helpful to think that the sections are there to foreground political, social and cultural perspectives, but that they are dimensions that together form the outcomes of educational experiences and that we made it our task disentangling them.

Three 'in-between' chapters that can be found before each section discuss and problematise what can be found in the chapters to come and use them like a stepping stone to show that the discussion continues. The book concludes with an afterword chapter by Christina Siry, who as a member of the *Journal of Cultural Studies of Science Education* community reflects on the troubling issues raised in this book and the contribution it presents.

We hope that by bringing these chapters together, they may encourage others to continue with such critical examinations, dialogues and debates regarding these issues.

References

Alton-Lee, A. (2003). *Quality teaching for diverse students in schooling: Best evidence synthesis.* Wellington: Ministry of Education.

Lin, T.-C., Lin, T.-J., & Tsai, C.-C. (2014). Research trends in science education from 2008 to 2012: A systematic content analysis of publications in selected journals. *International Journal of Science Education, 36*(8), 1346–1372. https://doi.org/10.1080/09500693.2013.864428.

Serder, M., & Ideland, M. (2016). PISA truth effects: The construction of low performance. *Discourse: Studies in the Cultural Politics of Education, 37*(3), 341–357. https://doi.org/10.1080/01596306.2015.1025039.

Chapter 2
Becoming of a Book

Kathrin Otrel-Cass

K. Otrel-Cass (✉)
Department of Learning and Philosophy, Aalborg University, Aalborg, Denmark
e-mail: cass@learning.aau.dk

K. Otrel-Cass et al. (eds.), *Cultural, Social, and Political Perspectives in Science Education*, Cultural Studies of Science Education 15,
DOI 10.1007/978-3-319-61191-4_2

BECOMING OF A BOOK
BY KATHRIN OTREL-CASS
IT WAS A COLD WINTER MORNING IN A SLEEPY TOWN IN SOUTHERN SWEDEN.....
....WE HAD ARRIVED FROM DIFFERENT NORDIC TOWNS AND CITIES, TO MEET IN AN OLD SEASIDE HOTEL..
...TO DISCUSS OUR FUTURE PLANS.

DURING OUR TIME AS A RESEARCH NETWORK WE HAD COME TOGETHER MANY TIMES TO DISCUSS HOW TO BE MORE CRITICAL IN SCIENCE EDUCATION RESEARCH.
LET US TRY AND ORGANISE THE IDEAS WE HAVE WORKED WITH
I THINK WE ARE GOING SOMEWHERE
WE ORGANISED AND REORGANISED THE IDEAS AND PROBLEMS WE HAD TRIED TO ADDRESS.
BUT WHERE COULD WE TAKE THINGS FROM HERE?

WE REALISED THAT THE WORK WE HAD DONE WASN'T FINISHED...
I WONDER HOW OUR SHARED INTERESTS COULD CONTINUE
THIS HAS BEEN GREAT
MAYBE WE COULD WRITE A BOOK TOGETHER?
YES BUT LETS MAKE SURE WE WILL STAY TRUE TO OUR IDEALS
WE WANTED TO MAKE A DIFFERENT BOOK, MORE CRITICAL, MORE POLITICAL, MORE SOCIAL AND MORE CULTURAL ...
...AND WE WANTED TO TROUBLE CONVENTIONAL THINKING.
IT HAD TO BE A PROJECT WHERE WE WOULD TO CONTINUE LISTENING TO EACH OTHER AND GO THROUGH THE WRITING AND THINKING PROCESS AS A COMMUNITY OF CRITICAL FRIENDS...

OUR PLAN WAS TO WRITE AS A COMMUNITY WHICH MEANT ALSO TAKING COLLECTIVE RESPONSIBILITY FOR THE FRAMING AND DIRECTION OF THE BOOK....
I THINK THE INTRODUCTORY TEXT SHOULD CAPTURE THE THREE THEMES SO THEY DON'T APPEAR BY MAGIC
SO WE SHOULD REALLY START WHERE WE TALK ABOUT THE CURRENT STATE OF RESEARCH AND THEN MOVE ON TO CHARACTERISE OUR THREE THEMES...
...YES BUT WITHOUT CREATING A STRAW MAN.
WE ALSO NEED TO BROADEN OUR IDEAS ABOUT THE READER...THEY COULD BE STUDENTS, TEACHERS, RESEARCHERS ...

DO YOU THINK WE CAN ONLY BE IN EITHER THE CULTURAL, SOCIAL OR POLITICAL THEME...
...AREN'T THEY ALL INTERTWINED AND WE FOREGROUND DIFFERENT THEMES THROUGH OUR WRITING?
FOR US IN THE CULTURAL GROUP WE FOUND THAT OUR STARTING POINT IS THE SCIENTIFIC INQUIRY.
...I LEARNED A LOT FROM READING THROUGH THE TEXTS OF THE OTHERS... I COULD SEE ALSO WHERE TO MAKE CROSS REFERENCES
BUT WHAT IS THE REASON FOR US ALL COMING TOGETHER IF WE THEN SEPARATE INTO DIFFERENT PARTS?
...I THINK WE NEED TO POINT OUT OUR OVERARCHING CONNECTIONS THEN WE CAN DETAIL THE NESTED INTERESTS AND CONCERNS..

MANY HOURS FOLLOWED OF TALKING AND THINKING...
... WRITING AND REFLECTING ....ON OUR OWN AND TOGETHER
...ONE OF THE IMPORTANT THINGS WE AGREED ON...
...WAS THAT THIS BOOK SHOULD BE DIFFERENT.
THE END

Chapter 3
In-Between Chapter: The Culture of School Science Inquiry Put Under the Microscope

Lotta Leden and Jonna Wiblom

Imagine a scientist! The bold spectacled man (yes, of course it's a man) – his forehead frowned and the white stained lab coat as a protecting shield – is shrouded in mystery and enigma. The lab coat scientist is the ultimate image of "ideal" science where knowledge is unproblematically accumulated through examining the natural world. A symbol that serves as an archetype for the construction of a school science culture where black is black, white is white and shades of grey refers to nothing but a poorly written book. School science culture, as it is commonly described, has its foundations in ready-made facts and "the scientific method." It seeks to provide insights into a canon of essential scientific phenomena and enculture students through pre-professional science training. Such culture takes on Thomas Kuhn's (2012) "normal science" through "normal science teaching."

Entering the chemistry, physics or biology lab, students are invited to participate in a mute game of *Jeopardy!*; neither do they ask the questions of inquiry nor do they answer them. As the show goes on, the sharp distinction between producing and *re*producing facts plays out as a mere chimera of the everyday puzzle-solving activity of "real" scientists. The out-of-school science consists of problems which are *believed*, and not known in advance, to have a solution. Despite the absence of student-derived seminal questions in science education, the practice of hypothesis testing, developing scientific knowledge, and raising students' interest in science are commonly mentioned purposes to justify the pilgrimage to the holy grail of science – the laboratory. All three chapters in this section take a point of departure in troubling school science culture through focusing the lenses towards *school science*

L. Leden (✉)
Kristianstad University, Kristianstad, Sweden
e-mail: lotta.leden@hkr.se

J. Wiblom
Stockholm University, Stockholm, Sweden
e-mail: jonna.wiblom@mnd.su.se

K. Otrel-Cass et al. (eds.), *Cultural, Social, and Political Perspectives in Science Education*, Cultural Studies of Science Education 15,
DOI 10.1007/978-3-319-61191-4_3

inquiry – the lead star of "normal science teaching." The authors twist and turn inquiry in its existing forms in school science culture, thereby scrutinizing prevailing images of science.

Lars Bang asks how school science inquiry through John Dewey and Joseph Schwab came to be. Referring back to Gilles Deleuze and Baruch Spinoza, he troubles the *ideal status* of *inquiry itself* and some of its specific elements such as the practice of asking questions – a false illusion of exploring or a search for genuine answers? Anna Jobér and Gerd Johansson trouble inquiry by making space and matter matter. Through studying the intertwined interactions of humans and nonhumans, the material is brought to life and given agency on its own. Practices in which artifacts are used and space is produced and organized are described as socially constructed, taken for granted yet difficult to grasp and describe. By illuminating such practices, seldom scrutinized in science education research, the authors direct our attention towards how school science is staged and how artifacts and space can be used as means to promote or inhibit agency, critical thinking, participation, and meaning making in science education. Paradoxical as it might seem, Jobér takes a point of departure in the microscope to move us beyond the level of micro cellular details. Under her illuminated lens, dimensions of space and its consequences for school science inquiry emerge to the reader. Johansson questions the spatial separation of practical activities (by the sink) and meaning making (at the desk) that reinforces a separation of body and mind, the doing and the thinking in the science lab. In this section, the identification and questioning of such dichotomies present in school science is the silver thread of meaning in all three chapters.

Bang builds on Deleuzes' concept of dramatization to expand the common question of *what* school science inquiry is to also include questions addressing *who*, *how*, and *when*. The way inquiry is emphasized in, for example, the Swedish national curriculum suggests answers to the *why?* as developing students "ability to plan, carry out, interpret and report experiments and observations, and also the ability to handle /…/ equipment" (Skolverket 2011). A search on "laboration" (Swedish for lab-work) on the Internet generates thousands of images of students, in too large or outgrown white lab coats, holding liquid-filled test tubes. Absent, however, are pictures of out-of-school lab-work in "real" scientific settings (they are to be found when searching for "laboratory"). So what are the "real" scientists up to if not lab-work? Producing new knowledge? Learning something? In line with the curriculum, the images that emerge on the laptop screen are an inquiry culture that appears to be all about the "doings" rather than the "knowings." In the center of attention is the systematic laboratory performance per se, step by step orchestrated by the clear and unquestionable sheet of instructions. Thus, school science culture in itself rests upon problematic images of science where a differentiation in "epistemic cultures" (Knorr-Cetina 1999) becomes invisible. Is school science about enculturing scientists-to-be into the culture of real science through learning "the scientific method" and juggle about with semi-scientific artifacts and accurate measuring? If so, the scientist in-the-making is at risk of being mighty disappointed by the messy mangling of "real science"; and the outsiders, the ones swimming besides the pipeline, might as well continue swimming. Is it a practice where the intentions are to replicate real science in a simplified form? Or, is it perhaps about raising students' interest in science through "learning by doing"? In the latter case, one might wonder *who* will be interested, in *what* exactly, and *why*? By troubling the culture of school science inquiry through dramatization, hybrid space, and socio-material perspectives, the focus is here and now yet connected to the contingency of historical and future global movement.

The paradigm of normal science teaching can be seen as co-constructed in a network of politics, curricula, teachers, students, teaching materials, facilities, spaces, and artifacts. In Leden et al. (2015), we encounter teachers who trouble, yet guard and co-create normal school science teaching. Here the practice of "facts and lab-work" constitutes a safe haven for teachers' identities – well-known structures are preserved and insecurity and messiness are avoided. Operating through the handbook of normal science teaching, only minor shifts will be made – far from the verge of challenging the core values of science education. Scrutinizing the culture of school science inquiry might only mean scraping on the surface of traditional school science teaching. What then, if acceptance proceeds change? Would admitting science education as a culture in its own right imply letting go of "mini-science" practice and close the fireproof doors to the holy grail of science education? What competences and what contexts become relevant when perspectives on science education shift from a culture of learning from inside out to outside in? From pre-professional science training to the fostering of "competent outsiders" (Feinstein 2011)? The decisions to be made for a science education will have the character of dealing with questions of "didactic transposition" (Achiam 2014) and "alchemy" (Popkewitz 2004). Meaning that decisions about *what? who? how?* and *when?* from real science cultures must be transformed and renegotiated into science education cultures. What would best serve students whom we want to become competent outsiders, "nonscientists who can access, interpret and produce the science most relevant to their lives" (Feinstein, Allen and Jenkins 2013, p. 314)?

In order to reform the culture of science education, we might very well need to expand our focus, from the inquiry-related "doings" and procedures to the inclusion of the joint meaning making of the scientific "knowings" relevant for participation in society. If we want science education to prepare students to deal with the diversity of science in everyday life, navigate media, and scrutinize conflicting arguments and awareness of marketing purposes, we might even have to leave school science inquiry (at least as we know it) behind. No matter how much we trouble it back and forth, is school science inquiry even something worth saving by troubling?

The tools for troubling used by the authors in this book section, dramatization, socio-materiality, and hybrid space, might constitute important tools in the work of scrutinizing various dimensions of school science cultures, such as teaching science for citizenship. No cultural manifestations of school science are exempted from the need of being troubled regarding agency, critical thinking, participation and meaning making. How would space and matter really matter in relation to a science for citizenship – and would the microscope still have a role to play?

References

Achiam, M. (2014). *Didactic transposition: From theoretical notion to research programme*. Paper presented at ESERA summer school, Neveshir, Turkey.

Feinstein, N. (2011). Salvaging science literacy. *Science Education, 95*(1), 168–185.

Feinstein, N. W., Allen, S., & Jenkins, E. (2013). Outside the pipeline: Reimagining science education for nonscientists. *Science, 340*(6130), 314–317.

Knorr-Cetina, K. (1999). *Epistemic cultures: How the sciences make knowledge*. Cambridge, MA: Harvard University Press.

Kuhn, T. S. (2012). *The structure of scientific revolutions*. Chicago: The University of Chicago Press.

Leden, L., Hansson, L., Redfors, A., & Ideland, M. (2015). Teachers' ways of talking about nature of science and its teaching. *Science & Education, 24*(9–10), 1141–1172.

Popkewitz, T. (2004). The alchemy of the mathematics curriculum: Inscriptions and the fabrication of the child. *American Educational Research Journal, 41*(1), 3–34.

Skolverket. (2011). *Curriculum for the compulsory school system, the pre-school class and the leisure-time centre 2011*. Stockholm: Swedish National Agency for Education.

Chapter 4
Education Extended: A Sociomaterialist Perspective on Science Education

Anna Jobér

4.1 Seeking New Perspectives

There have been several attempts to chart the exact nature of science education and scientific inquiry in the past couple of decades (see, e.g. Lundin and Lindahl 2014). This essay is no exception, for it seeks new ways of approaching and understanding scientific inquiry using a sociomaterialist perspective. In what follows I use sociomaterialism as a theoretical framework with which to examine a very specific element in school science culture: the use of laboratory work and the extensive use of objects, materials and things in science education. The sociomaterialist perspective is informed by recent post-humanist and post-structural currents in a range of disciplines, but until now it has not been commonly applied in science educational research. In this essay it is used to obtain new insights into the culture of school science, a culture that it has been claimed (Fenwick et al. 2015; Rudoph 2012) has been generally under-researched, with scant attention in the literature to the heavy reliance on objects, materials and things and how their extensive use permeates science education. I would argue that this is crucial, since the culture of science education by definition comprises practices, discourses and materialities, which over time express the continuities and discontinuities in custom and belief in the classroom and indeed in society at large. I hold school science practices to be a culture with patterns that are maintained and recreated, although this culture is often invisible at first glance because of its ubiquity and, arguably, its mundanity. The sociomaterialist view is also used to chart the complex networks of practices and accumulated knowledge and ideas that are transmitted, manifested and created by the same school science culture.

A. Jobér (✉)
Malmö University, Malmö, Sweden
e-mail: anna.jober@mah.se

K. Otrel-Cass et al. (eds.), *Cultural, Social, and Political Perspectives in Science Education*, Cultural Studies of Science Education 15,
DOI 10.1007/978-3-319-61191-4_4

Post-humanist scholars such as Donna Haraway and Karen Barad are the inspiration behind the theoretical framework of this essay. Each has theorized the concept of 'actors', where humans and non-humans alike jointly create and construct the world. By extension, then, my approach is informed by the theories developed by Bruno Latour and Annemarie Mol, among others. The theoretical framework thus challenges the notion that research and education are strictly human endeavours. To adopt the sociomaterialist perspective, to follow Annemarie Mol (2000), is thus to avoid 'me'—me as in a human being—as the all-knowing centre. This essay therefore highlights the importance of materiality in elaborating on the influence of human actions and beliefs on science education, and it considers materialities as a decisive component that influences educational outcomes, structures cultures and practices and yet also produces, stabilizes or destabilizes values, norms and figurations in the classroom and beyond.

My chosen focus is a common practice in science education, namely, the use of instruments, equipment, things, or objects in science classrooms as a means for pupils to learn and expand their knowledge. As a fictive case, I use a common classroom scene that many science educators will recognize, namely, a pupil using a microscope. Apply a sociomaterialist perspective, and this will be seen as a joint construct comprising a human (the pupil) and a non-human (the microscope). It is the figurations that this joint construct creates and performs that are the focal point here. It is instructive to compare this particular construct with the practice of writing. When writing, it is not solely the pen nor solely the hand that constitutes the action of writing; rather, it is what the writing does, creates and performs and how it is done that are the epistemic centre.

This practice—and, at the heart of this practice, the pupil and the microscope—has been chosen for a number of reasons. For example, as suggested by Tobias Röhl (2015), science teachers rely on a number of material objects in their teaching, objects that are not only important in their teaching activities but that also embody disciplinary knowledge. For example, certain objects used in the science classroom such as microscopes, safety goggles and lab coats are also commonly used in science practices in universities, in professional laboratories and in 'real' science in general. These objects could therefore be said to embody not only knowledge but also what that scientific knowledge represents, as derived from the 'real' scientific disciplines. Like an old chair with a rare and unique history, things and objects have a provenance that affects their owner and exerts a certain significance.

Still, the science classroom (and the field of research concerned with it) is human-centred (Murdoch 1997), and materials and things that matter are often missing from accounts of educational processes (Fenwick et al. 2015). Materials become a mere backdrop for human action, neglecting the fact that 'the physical apparatus' (Rudolph 2012, p. 2) in the classroom communicates clear messages about science. This essay therefore not only considers materialities as decisive components that influence educational outcomes, it also asks for greater recognition that materiality actively creates and configures educational practices (Fenwick et al. 2015).

4.2 Assemblages of Actors

The sociomaterialist approach derives from broader post-humanist and post-structural orientations and is closely connected to, for example, actor–network theory and science and technology studies. The key figures in relation to present study are Latour (cf. 2005), Mol (2000), Barad (cf. 1998, 2003) and Tara Fenwick (2011). Sociomaterialism demands a networked view of reality and a treatment of assemblages of human and non-human actors that generate ideas, practices and facts. One of the primary ontological priorities is that things and matter are constituted through their relations to or with other actors. To take one example, a mobile phone cannot be understood by itself but rather how it is used, what is does (or not) and how it is connected with satellites, human beings, etc. How a mobile phone can be used depends on the user, and the user is to some extent changed by holding a mobile phone in her hand (Lindström and Ståhl 2014). Indeed, it amounts to a 'dance of agency' (Pickering 1995, p. 22) where everything that happens is the result of the different actors' conformation or resistance (Serder 2015). Of particular interest, then, is the ontological supposition that ideas, practices, facts and the like are effects not of one thinker, nor of a specific action or actor, rather of assemblages and webs of relations. The actors in any assemblage can be human (e.g. pupil, teacher) or non-human (pencil, computer, microscope, curriculum); either way, they can be the co-creators of ideas, practices and facts (Gunnarsson 2015; Fenwick et al. 2015).

What should not be forgotten in the sociomaterialist approach is that assemblages and networks are not seen in present tense, in the here and now, nor in a static, one-dimensional understanding. The approach instead emphasizes assemblages and effects of assemblages that travel through time and space. It resembles the principle of provenance and how this provenance exerts power, leveraging cultural capital or understandings back and forth in time and space. This speaks to an understanding of how and why ideas, practices and facts are stabilized and destabilized. With the sociomaterialist perspective, it is thus possible to acknowledge 'the local practices of teachers and learners seriously while still accounting for practices that extend beyond here and now of a given situation' (Röhl 2015, p. 121). A particular classroom or practice is in this perspective connected to historical and global movements and other sites of ideas, practices and facts.

The practice of using objects, instrument and the like in the science classroom should not be thought a passive practice. It requires that things and matter act *together* with human actors such as the pupil or the teacher. Education in this sense is not only a practice by which humans interact with one another; it is a culture and a sociomaterial enterprise that relies on a number of material objects. Things and object are links in the historical and discursive chains of science education, and matter becomes an actor in the world's becoming (Barad 2003). I would not argue that every actor has the same capacity to act nor that agency is evenly distributed among actors, however; rather, agency is to adopt a standpoint that does not impose a human or anthropocentric perspective a priori (Latour 2005).

The sociomaterial perspective thus enforces a shift in the educational processes of meaning and agency. This opens up not only for other understandings of education; it also pushes the researcher into considering notable conundrums and non-linear ways of working. With that in mind, I now turn to the methodological issues in a sociomaterialist perspective and how best to tackle them.

4.3 Silent Objects

Fenwick et al. write of 'sociomaterial traces' that present them 'with a theoretical and methodological conundrum: objects themselves cannot speak back and explain their intentions' (2015, p. 127). What Fenwick and her co-writers are saying is that non-human actors such as objects and instruments are at first glance silent and cannot be asked questions or express an opinion. Ultimately, the researcher is left with a methodological conundrum when researching human and non-human actors. The sociomaterial perspective then requires the researcher to think of assemblages that stretch over time and space without any linear understanding of cause and effect. With a sociomaterial perspective, therefore, there are no simple solutions or straightforward methodological approaches to education. In this situation, it would be all too easy merely to acknowledge that the material *matters*; however, following Barad, I consider it worthwhile setting my sights on *how* the material matters and elaborate on how things might relate to the activity of thinking (Mol 2000). I can but ask myself how the joint human and non-human construct be understood and researched. How do non-humans and humans (or in the present case, microscope and pupil) act together? What is performed and figured through this practice?

I would argue that in order to elaborate on such questions, one has to apply an analytical procedure that includes tracing actors and actions, so following threads and movements that together reveal how the matter is jointly constructed and what figurations are produced (Gunnarsson 2015). One way of showing and tracing figurations is by telling stories (Fenwick et al. 2015) in order to encourage the researcher to ask questions about intellectuality as a more practical manner (Mol 2000). With the help of three concepts and a typical, if fictive, case (the pupil and the microscope), I will trace figurations to illuminate how a sociomaterial perspective can extend our understanding of education and the extent to which this perspective comes down to practices of knowing (Mol 2000).

With a sociomaterialist perspective, each actor is entangled with a myriad other actors in assemblages. Everything is connected in all possible directions, through time and space, in webs and nodes. In order to elaborate on a specific practice and to describe it in a comprehensible way, I follow Barad (2003) by performing an *agential cut* in order to look at a specific joint construct (the pupil and the microscope) that will be taken as the epistemological unit. The agential cut is determined by the researcher and the research apparatus and chosen to shed light on particular phenomena as meaningful. It is therefore important to note that these agential cuts create boundaries that define and limit the world, the research and what can be seen and understood.

In applying the sociomaterialist perspective, three concepts are of prime importance here: figuration, apparatus and phenomenon. The understanding and use of these notions derive mainly from Barad (1998, 2003), Haraway (1987) and Karin Gunnarsson (2015), albeit interpreted through the agential cut.

A *figuration* is a performative image, action or doing that exerts a certain effect (compared to provenance, for example) (Gunnarsson 2015). It can be a practice, discourse, norm, value, etc. A figuration is thus a performative action, doing or image that stabilizes and or is destabilized but is also inhabited; indeed, Kristina Lindström and Åsa Ståhl (2014) continue by describing it as a kind of verbal or visual figure that becomes occupied and sometimes reserved for a certain norm, value, practice, etc. An example from a feminist perspective is 'the hysterical woman', a figuration that is continually stabilized and destabilized through practices and discourses throughout history. Figurations can be traced through an apparatus where figurations are created, stabilized and destabilized (Gunnarsson 2015). The figurations in the present research can for that reason be seen as the result of iterative actions and practices in a science classroom culture, traced through an apparatus with phenomena. The question is what kind of figuration might be found in a science classroom? When one applies a sociomaterialist perspective, what could be said to be figured and embodied in science education? Is it, for example, the scientist as a white, middle-class man wearing a lab coat and rimless glasses? Is it a figuration that is inscribed into the school science culture, influencing who you should be or how you should act?

The *apparatus* can be said to be a field or mechanism of sorts, a site where figurations are created, performed and formed (Gunnarsson 2015); in the present case, the apparatus is *science education* and all that comes with such an assemblage, including teachers, pupils, materials and curricula. The science education apparatus is thus understood as an apparatus constituted through a number of practices, discourses, norms and values that are repeated across time and space, throughout science classrooms all over the world. It can be anything from inquiry-based learning to laboratory work to ways of learning and talking.

In an apparatus, practices are connected together and can be denoted phenomena. The *phenomenon* in the present study is a particular practice in science education, meaning the use of laboratory work and equipment as a way of learning, exploring the world and expanding one's knowledge. As a typical case, I have chosen to use a combination of pupil and microscope to illuminate the phenomenon. The focus is thus on the human (the pupil) and the non-human (the microscope) in a joint construct that stands for the phenomenon in the present paper. I agree with Barad that each phenomenon can be understood as human/non-human practice entangled in a web of discourses, sociohistorical legacies, norms, etc. that could be found in the apparatus.

4.3.1 Understanding School Science Culture

Two intertwined layers will be used to shed light on the main themes of the essay: *figuring authenticity* and *figuring ambiguities*. I have not set out to give a detailed map, nor yet full descriptions, but rather glimpses of the threads in a much larger web.

4.4 Figuring Authenticity

According to Vassiliki Zogza and Marida Ergazaki (2013), an integral part of real science—science as an academic subject—is inquiry as a process of constructing knowledge about the world. Other typical features are laboratory work, the scientific method and scientific instruments (Lederman 2008; Osborne and Dillon 2008; Rocard et al. 2007; Skolverket 2011). As such, science activities and practices in the classroom can in many respects be seen as deriving from academic science. Moreover, as Lundin and Lindahl (2014) argue, science teachers use these kinds of practices in order to present a more truthful view of scientific inquiry, to prompt interest and to lend authenticity to the science they teach by engaging in hands-on activities. As Heidi Carlone (2003) notes when discussing the notion of prototypical science, even though laboratory work is an important feature of school science, it projects school science as a prototype of 'real', authentic science.

In the last decade, however, critical voices have questioned the possibility of engaging in real science in school (Lundin and Lundahl 2014; Munby et al. 2000; Zogza and Ergazaki 2013). The character of school science in relation to the academic subjects of physics, chemistry and biology has been discussed (Carlone 2003; Munby et al. 2000). Magnus Hultén (2008), for example, maintains that school science has its own history and is subject to other forces—social and political influences, for example—in a way that professional science is not. One consequence, according to Hultén, is that the uniqueness of school science demands special attention and must be understood as a separate and unique activity driven by other objectives. It is in light of this discussion that the use of laboratory work as a way of learning science is of particular interest.

The notion that laboratory work is an essential part of science education can also be found as a societal discourse. A quick search for Google images, while not giving the full picture, does offer some interesting glimpses. Search on 'school science' and a good half of the first 50 images will probably be of someone wearing a lab coat. Search on 'science education' and again many of the first 50 pictures will contain a microscope. It would thus seem that the practical, inquiry-based dimension is a discourse highlighted both in and outside science education, a discourse that originates from laboratories and science practices in universities and companies, i.e. what has been called *real science* by some researchers. In addition, it seems like one object, the microscope, is a central feature in this discourse.

Thomas Popkewitz claims that '[p]edagogy translates academic knowledge into the world of schooling' (2004, p. 4) through a process similar to alchemy. Popkewitz argues that this alchemy process is a necessary part of schooling, pupils are 'neither mathematicians nor historians, translation tools are needed' (p. 4). This process is according to Popkewitz 'achieved through an assemblage of inscription devices that translate and order school subjects' (p. 4). The alchemy process might be necessary, but there are a number of pitfalls in this process in the science classroom (Lundin and Lindahl 2014) where the practices, or in this case, the pupil and the microscope as a phenomenon becomes on of all the phenomenon, or inspired by Popkewitz (2004) words, inscription devices, that configures the world. In this process, laboratory work becomes one of the important ingredients that projects school science as a prototype of 'real', authentic science, establishing sociohistorical legacies and as Carlone states 'taken for granted notions and sociohistorical legacies of science and science education that comprises the alienating nature of school science' (2003, p. 308).

In the science education apparatus, then, laboratory work and above all the microscope are important. When using an object or a thing (in this case, the microscope out of the pairing of the pupil and the microscope), it becomes something more than just a practical fulfilment of the curriculum. The pupil and the microscope together become an important part of the phenomenon that performs science education as a prototypical apparatus and risking adopting the assumptions that follow with the sociohistorical legacy of real science. I would argue that through laboratory work, the accepted granted norms and values become universal and are transported (Mol 2000) into the science classroom, just as the norms and values of what to be, how to act and who is preferred in science are established through the phenomenon. By placing the pupil in front of the microscope, the teacher recreates a historical, discursive chain that reinforces certain modes of action. The pupil and the microscope become an inscriptive device—a performative practice that over time and space repeatedly stabilizes figures and performs what science (or at least, what prototypical science) is and should do. What seems to be an innocent and accepted way of working in a local classroom can be linked with thoughts, actors, and discourses, however distant, which together produce iterative figurations. This perspective therefore encourages a deeper reflection on practices in the classroom. It also encourages researchers and teachers to remember that pupils are not scientist per se and, far from translation tools being all that is needed to negotiate authenticity (Popkewitz 2004), what pupils figure and perform must be recognized as the crux of all education.

4.5 Figuring Ambiguities

Laboratory work, then, is a common practice in science education. One of those practices is the use of microscopes to explore and understand more about the world. Some schools buy microscopes from private companies that specialize in laboratory equipment for schools and education. Following this thread, one can trace the object—the microscope—back to the manufacturer which developed and produced it (Röhl 2015). When following the traces left by the phenomenon out from the classroom towards the manufacturer's market, a number of questions arise. For example, private companies do not need to adhere to the national curriculum. Understandably, many companies do try to work according to the national curriculum and thereby support teachers; however, I would suggest that between school and commercial interests, there is a gap when it comes to what should be learnt and how it should happen. For example, a private company might have an agenda that differs from that of the school or their emphasis on how to use their products might be different. Röhl (2015) argues likewise that manufacturers try to construct desirable products by visualizing possible scenarios of how to use the object and how this might benefit learning. This can have a number of implications, of which I will concentrate on two in particular.

Firstly, when studying the Swedish advice website 'Allt om mikroskop' (lit. 'Everything about microscopes', www.mikroskop.se), it is clear that what they emphasize is the function. It is the function of the microscope that is the focal point, not the learning. A quick yet revealing search on YouTube for 'microscope', 'middle school' and 'pupils' produces much the same results: function, safety and how to use a microscope are the focal points of the video clips. This emphasis might not be a problem, but it is often different from what is stipulated in the Swedish curriculum and from what the teacher actually needs in the classroom.

Secondly, the objects used in schools are often simplified versions of 'real' scientific instruments, having been pared down to match schools' projected education needs. These objects are so simplified in order to avoid ambiguities in the classroom (Röhl 2015). Following the same argument, I would argue that this emphasis ensures that school science is a prototypical practice. Moreover, it reduces what actually could be seen and done in the science classroom. A similar argument has been put forward by Popkewitz regarding problem-solving practices in mathematics:

> The irony of this pedagogical practice is that the 'uncertain' and 'ubiquitous future' that the mathematical standards refer to is in fact not so uncertain or ubiquitous after all. The 'ubiquitous future' is fixed and ordered by the truth-telling practices embodied in the 'nature' and structure of conventional mathematics. Problem solving gives an illusion of flexibility, while the notions of nature and structure stabilize and regulate the uncertain future. At the same time, the inscriptions of the problem solver redefine the parameters of human agency, and what is open for scrutiny is circumscribed by the expertise that stabilizes and harmonizes the world of participation. (2004, p. 19)

In other words, science educational practices, such as the practice of using microscopes in school, give an illusion of exploring an unknown future; however, in science education this becomes just one practice that stabilizes the alchemical process, for example, by translating academic knowledge into the world of schooling and underscoring it by the use of simplified instruments that in the long run might hinder more complex understandings and arguments.

From a sociomaterialist perspective, what kind of figuration is performed and created here? In the science education apparatus, a space has been designated for things and objects designed and manufactured by companies with their own agendas that remotely control the classroom. By that I mean that a number of companies and institutions 'indirectly govern the classroom by developing material objects and other material entities' (Röhl 2015, p. 128). The pupil is presented with a tool designed to satisfy other agendas, and the pupil and the microscope together become a phenomenon that produces and reproduces certain agendas. Thus what is figured here is an education that is remotely controlled—an education where pupils repeatedly risk a reduced and simplified learning environment. I believe that the immediate response to this should be a need to reflect on the consequences of these kinds of simplified educational settings.

Of course, it is rare that placing a pupil in front of a microscope is in any sense truly problematic, and it is not hard to think of many other things that would be more problematic. In addition, when looking at just one specific phenomenon in one classroom in one part of a much broader assemblage, it might very well appear innocent and with no intentional thought behind it. When put together in assemblages that repeatedly perform figurations, however, the picture changes. Moreover, there are a number of objects and things that, from a democratic point of view, could be deeply problematic if used in the classroom, prototypical or not, simplified or not. I would therefore maintain that this is an issue that bears further scrutiny. Education and science education are powerful apparatus (Jobér 2012), and the question is who governs the apparatus. From a Swedish perspective, the issues regarding governmentality in a changed educational market have increased. At the moment, anyone who wishes to start a company and enter the Swedish edu-market (Axelsson et al. 2016) is free to do so, and education has become an apparatus largely without control, served by companies with ambiguous agendas.

4.6 Things and Thinking: An Extended Education

In this essay, I have looked at science education as an apparatus with a phenomenon, meaning a certain practice that creates figurations with implications for thinking and learning. The aim has been to understand more about science education by the application of a sociomaterialist perspective, with the pupil and the microscope as the example.

As the arguments illustrate, the pupil and the microscope can be used as a case to shed light on laboratory work as an iterative phenomenon placed in discursive and historical chains. The analytical procedure, in which I have traced what could be seen when using a sociomaterialist perspective on science education, reveals figurations that are created repeatedly. What seems to be figured here is science education as a 'prototypical' apparatus that fully adopts the accepted notions that follow on the sociohistorical legacy of the 'real' apparatus. The pupil and the microscope becomes a performative event that repeatedly stabilizes figures and performs what science is and should do, regardless of place or time. With a nod to Popkewitz, phenomena in science education become inscriptive devices that translate and order norms, values and discourses. What seems to be an innocent and accepted phenomenon without further thought in a local classroom can be linked to thoughts, actors and discourses far away, both geographically and temporally, which together iteratively produce and stabilize powerful figurations.

What is also figured is the remote control of classrooms—classrooms governed through recurring performative events and actions, such as using certain practices. The question is whether this governing practice (Popkewitz 2008) reinforces the ordering of exclusion and gaps in accountability. For example, what happens to pupils who are repeatedly faced with simplified versions of objects, actions or meanings? Do these objects become, as Rudolph (2012) claims, standardized packages that merely cater to the teachers' least demands rather than creating workspaces and sites of experimentation and uncertainty?

The use of microscopes in classrooms and labs in order to see, measure and count is therefore not 'simply *mental* operations' (Mol 2000, p. 17) but rather *mental* and *practical* operations. Materialities are structuring forces—crucial components that frame practices and produce values, which in turn influence its members (Isling Poromaa 2015). In order to learn more about education, one has to leave the immediate site of the classroom and trace practices along threads and assemblages of actors. This includes seeing education from a perspective where '[t]hings circulate in a midst of connection, cultural histories and symbolic values, but they themselves also compel activity' (Fenwick and Edwards 2010, p. 7). The view of education as something strictly human needs to be interrogated, and education has to be seen as something greater than the sum of its parts. The things, objects, instruments, practices, etc. and what they perform have to be acknowledged in pedagogy and discussed in daily classroom practices, regardless of the subject and whatever instruments, object or things are used.

Things have provenance and exert it and, along with their owners, are put in a chronology of ownership and agency. Nevertheless, even though the apparatus reinforces sociohistorical legacies and certain figurations, agency is not obstructed. I would argue that agency cannot be designated as a feature of things, objects or practices but rather that 'agency is a matter of making iterative changes to particular practices' (Barad 1998, p. 112) and those practices rely on more than non-human actors. Possibilities for agency and action exist at every moment, in every classroom, and *through* matter, things or objects connected to different sites, nodes and webs, regardless of place or time. There are therefore possibilities for action, critical thought and agency when viewing education as an extended endeavour.

References

Axelsson, T., Ideland, M., Jobér, A. & Serder, M. (2016, August). *Helping hands: Exploring Scholl's external network.* Paper presented at the 2016 European conference on educational research, Dublin, Ireland.

Barad, K. (1998). Getting real: Technoscientific practices and the materialization of reality. *Differences, 10*(2), 87–128.

Barad, K. (2003). Posthumanist performativity: Toward an understanding of how matter comes to matter. *Journal of Women in Culture & Society, 28*(3), 801–831.
Carlone, H. (2003). Innovative science within and against a culture of 'achievement'. *Science Education, 87*(3), 307–328.
Fenwick, T. (2011). Reading educational reform with actor network theory: Fluid spaces, otherings, and ambivalences. *Educational Philosophy & Theory, 43*(1), 114–134.
Fenwick, T., & Edwards, R. (2010). *Actor–network theory in education*. London: Routledge.
Fenwick, T., Doyle, S., Michael, M., & Scoles, J. (2015). Matters of learning and education: Sociomaterial approaches in ethnographic research. In S. Bollig, M. Honig, S. Neumann, C. Seele, & C. (Eds.), *MultiPluriTrans in educational ethnography: Approaching the multimodality, plurality and translocality of educational realities* (pp. 141–162). Transcript Verlag: Bielefeld.
Gunnarsson, K. (2015). *Med önskan om kontroll: figurationer av hälsa i skolors hälsofrämjande arbete*. (Doctoral thesis). Stockholm: Stockholm University. http://urn.kb.se/resolve?urn=urn:nbn:se:su:diva-116008
Haraway, D. (1987). A manifesto for cyborgs: Science, technology, and socialist feminism in the 1980s. *Australian Feminist Studies, 2*(4), 1–42.
Hultén, M. (2008). *Naturens kanon: formering och förändring av innehållet i folkskolans och grundskolans naturvetenskap*. (Doctoral thesis). Stockholm: Stockholm University.
Isling Poromaa, P. (2015). The significance of materiality in shaping institutional habitus: Exploring dynamics preceding school effects. *British Journal of Sociology of Education*. doi:10.1080/01425692.2015.1093406. http://dx.doi.org/10.1080/01425692.2015.1093406.
Jobér, A. (2012). *Social class in science class*. (Doctoral thesis). Malmö: Lund University. http://hdl.handle.net/2043/14071.
Latour, B. (2005). *Reassembling the social: An introduction to actor–network theory*. Oxford: OUP.
Lederman, N. G. (2008). Nature of science: Past, present and future. In S. K. Abell & N. G. Lederman (Eds.), *Handbook of research on science education* (pp. 831–879). New York: Routledge.
Lindström, K. & Ståhl, Å. (2014). *Patchworking publics-in-the-making: Design, media and public engagement*. (Doctoral thesis). Malmö: Malmö University. http://dspace.mah.se/handle/2043/16093.
Lundin, M., & Lindahl, M. G. (2014). Negotiating the relevance of laboratory work: Safety, procedures and accuracy brought to the fore in science education. *Nordina, 10*(1), 32–45.
Mol, A. (2000). *Things and thinking: Some incorporations of intellectually*. Quest XIV (1–2). Available at http://michiel.ipower.com/Quest_2000_PDF_articles/Quest_14_mol.pdf. Accessed 17 May 2016.
Munby, H., Cunningham, M., & Lock, C. (2000). School science culture: A case study of barriers to developing professional knowledge. *Science Education, 84*(2), 193–211.
Murdoch, J. (1997). Inhuman/nonhuman/human: Actor–network theory and the prospects for a Nondualistic and symmetrical perspective on nature and society. *Environment and Planning, 15*(6), 731–756.
Osborne, J. & Dillon, J. (2008). *Science education in Europe: Critical reflections: A report to the Nuffield foundation*. Available at http://www.nuffieldfoundation.org/sites/default/files/Sci_Ed_in_Europe_Report_Final.pdf. Accessed 25 Aug 2014.
Pickering, A. (1995). *The mangle of practice: Time, agency and science*. Chicago: University of Chicago Press.
Popkewitz, T. (2004). The alchemy of the mathematics curriculum: Inscriptions and the fabrication of the child. *American Educational Research Journal, 41*(1), 3–34.
Popkewitz, T. (2008). *Cosmopolitanism and the age of school reform: Science, education, and making society by making the child*. New York: Routledge.

Rocard, M., Csermely, P., Jorde, D., Lenzen, D., Walberg-Henriksson, H. & Hemmo, V. (2007). *Science education now: A renewed pedagogy for the future of Europe*. Available at http://ec.europa.eu/research/sciencesociety/document_library/pdf_06/report-rocard-on-science-education_en.pdf. Accessed 27 May 2015.

Röhl, T. (2015). Transsituating education: Educational artefacts in the classroom and beyond. In S. Bollig, M. Honig, S. Neumann, C. Seele, & C. (Eds.), *MultiPluriTrans in educational ethnography: Approaching the multimodality, plurality and translocality of educational realities* (pp. 121–139). Transcript Verlag: Bielefeld.

Rudolph, J. L. (2012). Teaching materials and the fate of dynamic biology in American classrooms after sputnik. *Technology and Culture, 53*(1), 1–36. doi:10.1353/tech.2012.0037.

Serder, M. (2015). *Möten med PISA: Kunskapsmätning som samspel mellan elever och provuppgifter i och om naturvetenskap*. (Doctoral thesis). Malmö: Malmö University. https://dspace.mah.se/handle/2043/17966.

Skolverket. (2011). *Curriculum for the compulsory school, preschool class and the leisure-time centre* 2011. Available at www.skolverket.se/publikationer. Accessed 16 Dec 2011.

Zogza, V., & Ergazaki, M. (2013). Inquiry-based science education: Theory and praxis. *Review of Science, Mathematics & ICT Education, 7*(2), 3–8.

Chapter 5
The School Science Lab: Hybrid Space and the Production of School Science

Gerd Johansen

5.1 Introduction

This chapter seeks to disassemble the physical space of the school science lab; the walls and windows, the volumes of air, moveable furniture and fittings, as well as fixtures are the silent parts of school science. In science education research, the physical lab is conventionally just part of the context and is given little cause for concern. In their excellent review, "Learning in and from Science Laboratories", Hofstein and Kind (2011) provide an overview of previous and emerging themes related to science labs. However, they do not provide emerging research on the lab as a physical space; indeed, it seems like the lab is a connotation for the practice of doing science. The lab is so "common" that we do not see it. In the following, I turn this convention on its head, as I foreground the science lab itself. In so doing, my intention is to unsettle the naturalisation of the physical space and its contribution in constituting school science practice.

The physical space of the school science lab plays a part in the practices of school science. In order to explore these practices, I argue that the invisible and inaudible physical space needs to be disrupted, as it is part of what Heidi Carlone (2004, p. 411) describes as the "tenacity of the socio-historical legacy of school science". Because we can see the physical space of the lab, we think we comprehend it: this is what Henri Lefebvre (1991/1974) calls the illusion of transparency. The school science lab is not a "passive" or "neutral" contributor to the production of school science. The physical space provides possibilities and constraints for participation in inquiries and other teaching and learning activities, and therefore, it has a role to play when discussing what science education ought to be and how it should be enacted.

G. Johansen (✉)
Norwegian University of Life Sciences, Ås, Norway
e-mail: gerd.johansen@nmbu.no

K. Otrel-Cass et al. (eds.), *Cultural, Social, and Political Perspectives in Science Education*, Cultural Studies of Science Education 15,
DOI 10.1007/978-3-319-61191-4_5

The science lab is a social space – and as most spaces it is naturalised, taken for granted (Lefebvre 1991/1974). The school science lab is made by people and constantly remade by teachers and students: they leave their imprint on the physical environment. It is a social space, a social construction where there are relationships between perceptions, spatial practices and its representations (Lefebvre 1991/1974). Students and teachers' actions are neither completely restrained by the physical space of the lab, nor are their possibilities for actions limitless. Their spatial practices are related to how they perceive and "understand" the science lab. According to Edward Soja, this can be seen as a socio-spatial dialectic where "there exists a mutually influential and formative relation between the social and the spatial dimensions of human life, each shaping the other in similar ways" (Soja 2010, p. 4). Making this dialectic more visible is done in order to illuminate how space is a constitutive, yet rarely acknowledged, part of school science.

More specifically, this chapter draws on Lefebvre's (1991/1974) claim that a physical space is contradictory in the sense that it sends a range of different messages – some in conflict with each other. This way of perceiving space makes it challenging to approach the science lab analytically. A main part of the chapter's argument is that the school science lab is a particular kind of hybrid space, it is drawing on traditions from both science *and* school spaces. Arguably, all social spaces can be regarded as contested, drawing on a multitude of practices, perceptions and representations (Lefebvre 1991/1974), rendering them hybrid by default. *This* particular hybridity is creating an amalgam of science and school that is vital to unravel. The supposition of science lab as a hybrid space is a point of departure and will be explicated throughout this chapter.

This chapter explores the school science lab as a physical space while students are doing a closed inquiry. The students are given lab equipment and a traditional recipe for carrying out small practical chemistry inquiries, and they carry out their investigations within predefined areas of the lab. To address the socio-spatial dialectic between students' science practice and the physical space, the following question is asked:

- What science practices are produced within the school science lab?

The question is examined by employing a micro-ethnographic approach to explore one group of students during two closed inquiries.

The chapter starts out with a theoretical as well as an empirical argument for the school science lab's hybridity, followed by an empirical investigation into how science is produced in this hybrid space. The chapter ends with a discussion on the lab as a space for the production of school science.

5.2 Social Spaces: Their Organisation and Function

In this section I will argue that the school science lab is a hybrid space by drawing on literature describing professional science labs and school classrooms. I begin by explicating my approach to physical space as socially produced and as a site for production.

The school science lab is a man-made construction used for teaching and learning science, as well as the material objects used for this purpose. Lefebvre (1991/1974) argues that the social space is a space where the material and bodily actions "reside" – it is not an abstractly verbalised space. Moreover, social space is not a product among other products, but subsumes produced objects. Social space is produced and is a site for production, e.g. of school science content, and (ordinary) spatial practices, in this case the students' practice of school science. As a social space is in constant flux – it changes – it becomes difficult to pinpoint, difficult to describe, quite contrary to a naturalised view of physical (social) spaces as fixed:

> Does it make sense to speak of a "reading" of space? Yes and no. [...] Both natural and urban spaces are, if anything, "over-inscribed": everything therein resembles a rough draft, jumbled and self-contradictory. Rather than signs, what one encounters here are directions – multifarious and overlapping instructions. (Lefebvre 1991/1974, p. 142)

Even if the school science lab is more "ordered" than an urban city space, it is still characterised by multifarious and perhaps even self-contradictory elements. Moreover, Lefebvre's statement provides an opening for several possible approaches on how to deal analytically with space. One approach is described by Louise Ravelli (2008) wherein she builds on the multimodal social semiotic approach from Gunther Kress and van Leeuwen (2006). I will draw on this approach to some extent but mainly turn to Edward Soja who in his book, *Seeking Spatial Justice*, develops Lefebvre's ideas.

As previously mentioned, spatial practices can be approached as a socio-spatial dialectic where physical space and social activity continuously interact. In this chapter, three characteristics of physical space are seen as particularly important: the space's boundaries, its nodes or centres and its distances.

Classrooms and science labs are typically indoors, though this is not a strict requirement; class can be held in a field under a tree, and scientific experiments can be conducted in town squares. The walls of the room mark the boundaries that distinguish the inside from the outside – in this case school science from "non-science". However, rooms inside schools are, even if sectioned off, to an extent permeable; there are ventilation systems, windows and doors and the possibility to walk in or look out. Within the boundaries of this physical space, there will be different local centres or nodes:

> Our actions and activities will tend more or less to be nodal, focused around particular centres or agglomerations, and this centering or nodality will generate unevenly distributed advantages and disadvantages depending on location and accessibility with respect to centre or node. (Soja 2010, p. 72)

In the classroom such centres can be centres of attentions in different ways, such as a centre for some practical activity or for a theoretical activity. When students are dividing tasks, there might be some doing the "writing" and others doing the "working". Other centres might be created as they are given prominence. According to social semiotic theory, aspects such as size, colours or visual/auditive disruptions might act as focal points (Ravelli 2008). In a science lab, there are hubs: places of special importance. The sink is an example of one such place, as there is often a need for water, leaving the sink as a modern equivalent to the village well. Distance as a spatial feature is not equal to objectively measured distance, as the distance from a centre will depend on what the centre is and what is placed in the distance. For instance, a student can easily follow the classroom screen from the back row, but not be able to see his or her neighbour's computer screen, which may function as the group's theoretical centre. After this explication of the spatial characteristics – boundaries, centres and distances – I will relate them to aspects of social activity.

Drawing on Soja (2010), to have access to information, equipment and artefacts is important in order to partake in a practical activity in science. In doing science it is vital that information is shared, and if the point is to have hands-on experiences, these need to be had by all. In other words, location and distance are important. To be physically near the equipment might ensure control over practical activities, but distance is relative, because in some situations one might be further away and still have access to what is going on. Moreover, in the more theoretical part of the practice, access to information on screen might be important and it might not be sufficient to listen. During science practices, walking around is usually allowed or even deemed necessary, for instance, when walking to the sink to rinse out a beaker.

In the science practices, the distributed advantages and disadvantages provide consequences for participation and distribution of power (Soja 2010). There will be a question of whether it is possible to participate, e.g. if the information on screen is only visible to two out of four students – the four do not stand on equal terms. In his ethnographic work, *Tangled up in school*, Jan Nespor (1997) describes how the bodies of children are regulated and controlled in school spaces. There will also be differences in power between students, with the possibility to control and take charge of an activity varying. The same might be said in relation to having access to material and immaterial resources, as well as their distribution. Moreover, power in a classroom also relates, to a certain degree, to "cultural imperialism" – there are some (things) that are right and others that are not. There are thus norms that regulate students' science practices. In the empirical section, the dialectic between physical space and science learning as a social practice will be described.

5.2.1 Classrooms: Function(s), Organisation and Practices

The main function of the classroom is to facilitate practices of teaching and learning. Here it is argued that the physical space – the classroom – invokes the practices and how they are understood. Leander et al. (2010) argue that the classroom as container is the prevailing metaphor – a container in which some activities occurs. They continue:

> One might almost see the classroom as the epitome of immobility as well, representing not only conventions of material structure but also conventions of teaching schedule, of seating charts, and seatwork routines. (Leander et al. 2010, p. 332)

A part of the room's invisibility and perceived stasis is the embedded routines such as time slots for lectures and activities – in assigned areas in the building. Moreover, there are conventions and routines for where students sit and where and when they can move. Leander et al. (2010) continue their argument: that in order to change our view of teaching and learning, there is a need to approach the learning environment as a social space, to make changes in this space.

Gunther Kress et al. (2005) argue that displays and spatial organisation of classrooms is a feature of how subject matter, in their case English, is (re)produced. They argue that students' seating arrangements are partly connected to how teachers interpret and conduct both the curriculum and other guidelines for teaching. They analyse four different classrooms. In one of the classrooms, the teacher had substituted some of the traditional furniture with more informal furniture such as sofas. By doing so, in addition to choosing to display students work and popular media posters around the walls, the teacher produced a more informal notion of the space. In another classroom, the students were seated in groups in such a way that the teacher could see all the students when seated at his desk – a sort of panopticon. However, this teacher often walked along a semicircle at the front of the room, thereby standing behind several of the students. In this class the displays on the walls were "official" and hung there by the teacher, while at the back there were presentations of the students' work. They argue that this classroom produces another practice, one where the teacher is more in control. Prominence is given to teacher's displays as they are situated behind the teacher at the front of the room.

Quite a different classroom is described by Pauline Jones (2008). She describes a small community school where all the students share one room. However, to separate juniors from seniors, there are barriers, e.g. shelves. The teacher's desk is placed in the centre of the room, allowing the teacher to oversee most of the space. This organising of space does not produce a clear front or back. The desks are joined so that students work together, and moreover, the school building is a central part of community life. Jones (2008) argues that this classroom is a space that is homely and that this quality facilitates a more everyday approach to subject matter as school becomes a continuation of home life.

5.2.2 *The Professional Science Lab: Function(s), Organisation and Practices*

The professional science lab is a strikingly different sort of space than a homely small community school – or any "ordinary" school, for that matter. To highlight some aspects of the spatial features of the professional lab, I turn to writings based on two major ethnographic fieldworks. In her seminal work, *Epistemic Cultures*, Knorr Cetina (1999) describes two different types of science laboratory: a high-energy physics lab and a molecular biology lab. She argues that the role of a science lab is to relocate phenomena from where they occur naturally and to control them, e.g. through isolating variables as well as controlling when they happen. This constitutes a systematic relocation that works differently in the various fields of science. She argues that as the content is specific to each scientific field, the social practice, the culture and the spatial organisation are very different. The spatial arrangements of these two types of labs also have different functions in the production of new knowledge: they are part of different cultures. She argues that the collaborative traditions are different in these two settings. In the molecular biology lab, the researchers and technicians mostly worked alone – whereas experiments in high-energy physics involved a huge number of people.

As the lab that is the concern of this chapter shares more traits with a biology lab than a physics lab, it is appropriate to dig deeper into the spatial arrangement of a biology lab. I turn therefore to the book *Laboratory Life* by Latour and Woolgar (1986). They provide a description of the physical space that constitutes a scientific lab and the work that goes on inside this space. Although they are critiqued by Davis Baird (2004) for not giving materiality sufficient prominence, their work might serve as a starting point for a description of one type of lab (microbiology). In this lab there are designated areas for different purposes – and different people. The researchers are mainly located in the "desk area" and they deal with analytical messiness, whereas technicians deal with wet material and material messiness. Both technicians and researchers walk between different areas such as desks, benches, storage and apparatuses – and there are "detours" involving formal as well as informal talk. Latour and Woolgar (1986) further state that the material layout of the laboratory has been constructed based on apparatus. In other words, the equipment – the physical artefacts that are needed to make the necessary transformations of material and provide inscriptions vital to analysis and scientific argument – are part of the production of the lab as a social space. Apparatuses are part of the definition of the function of (separated) spaces, e.g. a nuclear magnetic resonance machine has its own space. Different apparatuses and artefacts hold prominence – they are a vital part of the production. Moreover, as writing is a significant part of the work for both technicians and researchers, it is important to have facilities for writing such as desks and chairs and technological tools for the actual writing process either pen and paper or computer.

5.2.3 Hybridity of School Science Labs

Obviously, classrooms and science labs have different functions. These differences will be manifested in which practices are in the centre of attention. One difference is that whereas professional science is directed at producing new knowledge, Gyllenpalm and Wickman (2011) found that among student teachers lab tasks were seen mainly as a pedagogical activity. This finding is also consistent with research findings in Ian Abrahams (2009) study of practical work in school science. In other words: the "problem" is not driving the investigation. School science is mainly concerned with reproducing through the use of a recipe for practical activities with a known outcome (Tiberghien et al. 2001). In some respects a professional lab work might have similar traits as there are strict procedures to follow in order to ensure quality (Latour and Woolgar 1986). The school science practices provide a rather distorted image of professional science. In a Norwegian context, Per Morten Kind (2003) found that secondary school students had very simplified notions on what it meant to conduct scientific experiments.

Another difference between school and professional science is the schemes for distribution of advantages, such as resources and possibilities to participate in the practices. In the professional lab, there will be different roles assigned to different people as the work is distributed according to experience and competence. In school science there is much less distribution of work as all students are to "learn the same". To conclude, school science deviates from professional science, and perhaps, this is connected to how the school science lab is laid out.

The school science lab will also double as an ordinary classroom. In such cases it is expected that the centre of activity is located towards the front (Leijon 2016), where the teacher's structure of activities is given prominence. In a traditional teaching situation, the teacher will have the power of exercising regulations and control over students and practice (Nespor 1997). Students' participation is regulated and specified during practical activities, but as there is greater disorder within the boundaries (because of e.g. students possibility to move more freely), students have the dis- or advantage of being less visible for the teacher. However, both teacher and students adapt to the space they are occupying (Jobér 2012).

The double function of the site for science lab and traditional theoretical teaching and learning activities sets an imprint on the physical space as it has to facilitate both functions. Moreover, as the space is designed in this way, it encourages some practices over others. How this plays out in one concrete classroom/lab will be explored next.

5.3 Material and Method

The problems at hand – (a) to substantiate empirically the argument of the science lab's hybridity and (b) to explore the mutual influence between the social and spatial dimensions of school science practice – pose several challenges. The first major challenge is to upend the presumption of humans as the sole centre for research attention when dealing with science inquiry. Here space is just as much in focus of the research as humans. This gives the following implications: firstly, in the descriptions students have no names, they are simply moving and doing. Secondly, what they say and their meaning-making are of little interest unless this is directly related to space. The second major challenge is what to look for and describe. Even such a confined space as the science lab provides very rich material. As the number of pages and the reader's patience are limited, there have to be delimitations of descriptions and these will rest upon the partiality of the researcher. From another research field, Fataneh Farahani (2010) has written on the problems of being an insider in ethnographic research. As I have been in many school science labs, there is a risk I provide descriptions of what is "obvious" to science teachers and peculiar to others. The third major challenge is, as mentioned in the introduction, that space sends different – and often conflicting – messages. It becomes even more complex when there is a science inquiry going on. A consequence of this is that it will at times be difficult to make straightforward interpretations. There will be conundrums and unanswered questions. However, I do not see it as the purpose of this text to provide clear answers but rather to raise an awareness of the dialectic between space and school science practice.

In the introduction it was argued that the lab is naturalised and, moreover, that students' interactions with the physical environment are naturalised. The socio-spatial dialectic is thus hard to describe. Furthermore, in order to capture the negotiations in the socio-spatial practice, it is important to describe the "static" lab as well as the lab with actions. The "lab with actions" includes artefacts and inquiry, but these aspects are not focused in this chapter.

To explore "the natural", a micro-ethnographic approach is chosen. This approach allows for detailed descriptions of practices and physical surroundings over short time spans. It provides possibilities for a moment-by-moment analysis of video recordings with respect to students' interaction (Erickson 2006), here with the focus on their interactions with regard to space. These interactions are complex, as they are not actions and reactions in a neat order but multiple actions occurring simultaneously. Moreover, each action remains elusive as there are several possibilities for interpretation (Erickson 2006). In addition, the micro-ethnographic allows for descriptions of the confined physical space: the "fixed" school science lab. In more traditional ethnographic approaches, one would emphasise a larger part of the surroundings to provide a rich description of the space and culture; see, for example, Hammersley and Atkinson (2007) on this point. Seeing the room both as "fixed" and "engaged" by students' actions involves a sort of reductionism, a simplification needed to reduce complexity for the purpose of making this a comprehensible text. When applying a micro-ethnographic approach, it is a challenge, according to Julia Snell, to weave the descriptions of short-spanned socio-spatial practice into "a bigger picture" (Snell 2011). This is a challenge not addressed in this chapter.

To reconstruct the "fixed" space for an analytic description, video recordings from several cameras were used: a stationary camera as well as portable cameras carried by students. Descriptions of students' socio-spatial practices are based on transcripts of footage from one of the portable cameras. The transcriptions and the subsequent descriptions are influenced by explicit as well as implicit theoretical perspectives on space and science practices (Erickson 2006). This particular video material is collected by another researcher and with another research purpose, namely, to capture students' verbalised reasoning. By reusing material one runs the risk of losing context (Andersson and Sørvik 2013); to mitigate this problem, I have worked with the complete video material from this class, comprising footage of several groups during different practical settings. From my experience as a science teacher educator, this is a quite normal science lab, although perhaps a bit tidier than most.

The micro-ethnographic requirement of detailed descriptions limits the number of such descriptions:

1. The fixed lab – a lab "without" time and action. This description seeks to substantiate the argument of hybridity as well as form a frame of reference for students' socio-spatial practice.
2. Students' engagement with disciplinary content within this social space (lab). In other words, the mutual influence between the social and the spatial dimensions of school science practice. Different students or groups of students and various activities could have served as the focus of the descriptions. However, I have chosen to focus on one group of five students during two interconnected situations to serve as the focal point in order to enable a description and discussion of the science practices that are produced.
   (a) The students are sitting at their desks working with the electrolysis equipment, computers and mobile phones. The description is based on 4 min of a practical investigation that spanned approximately 25 min. One of the students carried a portable camera.
   (b) The students work at the fume cupboard. The description is based on 3 min of a practical investigation that spanned approximately 10 min. One of the students carried a portable camera.

There is a fleeting transition between these situations, as the students move from one practical activity to the next. The situations are chosen as they deal with different areas in the lab: the school desk area and the fume cupboards area, and the transition between these. In other words, they act as two lenses for the lab's hybridity.

5.4 The School Science Lab and Science Practices

The school building is quite new; when this is written, it is not yet 10 years old. The "newness" stands out on the footage. There are less than 1000 students spread across vocational and general studies at this upper secondary school. When students move around the building with the portable camera strapped on, there are no visible dents, discolouring or traces of vandalism, neither inside nor outside. There are large windows to let daylight in and large common areas for mingling and curricular activities – such as a staircase that can serve as a theatre or concert hall. There are also windows inside the building – making visible or half-visible activities inside the classrooms. Most of the building and furniture are in black and white; however, there is some use of bright colours such as yellow and green on doors and some pieces of furniture. Then, we enter one of the science lab.

Fig. 5.1 The science lab – after a finished lesson

5.4.1 The Science Lab

The first impression is that this room provides "good working conditions" for students – the room seems tidy, spacious, light and airy. This room produces an atmosphere of brightness, openness and transparency. According to Leander et al. (2010), these can be seen as surface features that are commonly perceived as constituting a good teaching-learning environment. The forthcoming analytical description intends to probe and trouble this common notion.

The science lab looks "new" and is a rectangular room. It is made up almost entirely of straight lines (no curved objects or walls). The boundary is clearly defined, and the walls in the science lab are white with some bright yellow and green used as contrast colours, e.g. on the doors. There are windows on three of the walls. This makes the room appear "light" and "airy". One can stand in the outside corridor and look in, as well as sit inside and look out onto the corridor, making it permeable. Moreover, this means that passing students, teachers and principals have the possibility to watch what is going in inside the room. Outsiders have some visual control. Perhaps this can be part of regulating the students' bodies; they are aware they are always watched, so they need to behave accordingly. The "back" wall is glass from the waist up, providing a good view of outside "nature". From a pure science perspective, the form and placement of the windows imply that this room is not fit for experimentation with light.

The front wall has an electronic board as well as an analog whiteboard; see Fig. 5.1. There is, therefore, a clear division of front and back sections of the room. The official site of knowledge transmission is the front – with apparatuses installed for this purpose. In the back of the room, there are benches, fume cupboards and a door leading to the storage room. The back of the room is "hands-on" whereas the front is "mind-on". Interestingly, in the introduction to this practical work, the teacher chose to stand in the back while introducing equipment and in the front while presenting background theory. One might speculate that the teacher's spatial positioning contributes to the mental segregation of "think" and "do", so often referred to in science education literature (see, e.g. Hofstein and Kind (2011)). If space plays a part in the division between "do" and "think", then what is given prominence is usually the front, i.e. theory.

The teacher's desk is in the front, even if it is not in the centre of the room it becomes a kind of hub. It is from here much of the activities in the room are directed. The distance to this hub becomes a choice to make for students. Moreover, the teacher's desk is flat and immoveable. The desk has a built-in sink but no fume system, making it fit to conduct various scientific demonstrations. As it has a fixed position, it probably amplifies the perception of the front as the centre of attention, requiring more effort to relocate a new centre of attention. The students' desks are flat, moveable and not connected to water and power. Moreover, the students' desks are double width so they can sit in pairs. When the lesson is over, the students tidy up and move the chairs and desks into rows; see Fig. 5.1. During the lesson this day, two and two desks were fitted together to form group tables – leaving some students with their backs to the front while others face the front of the room. However, there are swivel chairs so students can easily rotate to follow the teacher – but then, they might not have access to their writing equipment any longer. The moveable furniture for students provides a higher degree of freedom as to how to organise the room – and thus, the students' activities. Moreover, this flexibility in placing furniture produces the possibility to create different pathways through the room.

The lab is fitted with cupboards, and in the back of the room, cupboards are arranged kitchen-style with a sink. Along one of the sidewalls, there are four sinks and a rack for drying glassware. Beside the sinks there is a stand for lab coats and a first aid kit. Beside the door there are two fume cupboards with glass shields to pull down, each approximately 1 m wide. The combination of these items is a marker of the function of this space – it gives the room an air of "scientificness". None of these material artefacts are hidden away, e.g. by sliding doors, however, nor are they placed in the centre of the room – they are discretely placed along the walls and not given prominence. One might speculate if this choice is deliberate to downplay the amount of "scientificness", the lab could otherwise have had if these markers were more prominent.

There are some material artefacts on display. Firstly, there are several charts over various living organisms (different birds and bees): the most prominent of these are two life-sized charts of the human body. However, the human body is presented scientifically – they are not representations of the "natural" body. What is prominent is the human body as bones and muscles – the theoretical scientific body. In other circumstances this would be considered rather peculiar décor. Here it is part of the enculturation into science.

Other physical artefacts in the room are a model of the double helix and some green plants in the window at the back. The green plants are perhaps a part of another class' on-going experiment. All the items on display are from the realm of biology. The items on display all seem to be there by the teacher's initiation: students are perhaps not given access to the organisation of displays. The items are neatly organised along the walls. The space seems tidy and there are no broken up displays or cluttered areas. This is a room used for general science – and this day it was used for teaching and learning electrochemistry. The items on display signal implicitly what type of content the teachers identify as important in science. It is well worth noticing – and troubling – the choice of décor, as it plays a part of shaping students' view of science. The function of the science lab is teaching and learning science. So then the question becomes: what science? And another question: how is this science staged? As this lab's displays are mostly items from biology disciplines, it would be unfair to assess it in terms of a physics lab. However, if the students have lessons on physics or earth science themes in this room, it becomes a question if the décor works well with respect to disciplinary enculturation.

Moreover, this lab stands a stark contrast with the professional science labs – there are no very technical scientific instruments in sight. The most high-tech instruments are the smart board and sound system – in other words, instruments of school. The science lab-work apparatuses are few and placed along the walls – and they are not prominent. They are, however, not hidden. Thus, this room is not quite an ordinary classroom and not quite a science lab. One might argue that this hybridity – this mixture of two different social spaces – is necessary in light of the objective, education. Whether the balance between "school" and "science" is appropriate, and which one of the two is the most advantageous, is a question that cannot be answered without exploring the actual practice.

5.4.2 Science Practices

The analysis of a physical space in a school building might be interesting in itself, but here it is of interest to investigate the science lab as an accommodating as well as restricting factor in the production of students' science practices:

> Activity in space is restricted by that space; space "decides" what activity may occur, but even this "decision" has certain limits placed upon it. Space lays down the law because it implies a certain order – and hence also a certain disorder [...] Space commands bodies, prescribing or proscribing gestures, routes and distances to be covered. (Lefebvre 1991/1974, p. 143)

In this part of the chapter, what is particularly interesting is to explore how the physical space limits the practice. Space restricts the practices within, but it also provides possibilities. The science lab provides possibilities for students to slouch and sleep, walk around and engage in small talk, or my personal favourite, look out of the window – and dream. Students do all these things and more in the lab, but that is not the intention of school science.

5.4.2.1 The Context of Activity

The students enter the lab and are seated in groups of four or five. The teacher holds an introductory talk at the front of the room using the smart board projector. Then the teacher walks to the back of the lab to give instructions on proper use of the equipment which is placed on a trolley beside the teacher. The five students on whom this research focuses did several practical activities during this lesson. They could choose in what order they wanted to do the five activities with the teacher organising the access to the two fume cupboards, as there would be a queue otherwise. All the students got up and walked over to the coat stand to get coats and protective glasses. One of the students (Student B) fetched the equipment for the electrolysis activity and connected it. The first situation is occuring approximately 20 min after the introduction:

5.4.2.2 Activity Location: By the Desks

The five students are sitting around two (double) desks formed as a square at the very back of the lab. Pairs of students are facing each other, while the fifth sits at the head of the table. Student B has taken charge of the equipment and is carrying out the practical activity. Student B holds the electrolysis rods and has a fairly comfortable work position. This means that the equipment is out of reach off the two sitting at the other side of the table.

Table 5.1 and Fig. 5.2, respectively, show what the students are doing and where they are seated around the table. In Fig. 5.2, the equipment is indicated in front of Student B.

In this situation, the two students C and E are not using any artefacts related to the science activity. They are, however, at other times using material artefacts as part of their disciplinary actions. Moreover, Student E is also silent in this situation.

These students chose to sit at the back of the lab, as far away as possible from the centre. In the classroom the centre is most often located in the front – around the board and teacher's desk. It is quite possible that this was a deliberate decision, as this is usually the area where the teacher has the least amount of control. It is an area of perceived freedom and is also close to one of the windows facing the common area, enabling the students to oversee what was going on the outside as well as in

Table 5.1 Students' disciplinary actions at the table and their use of material artefacts for this purpose

Actor	Actions (disciplinary)	Material artefacts
Student A	Reads/writes on computer, talks	Computer and task sheet
Student B	Holds the rods and talks	Electrolysis equipment
Student C	Talks	
Student D	Takes pictures (talks)	Phone
Student E		

Fig. 5.2 The group's seating arrangements around the table

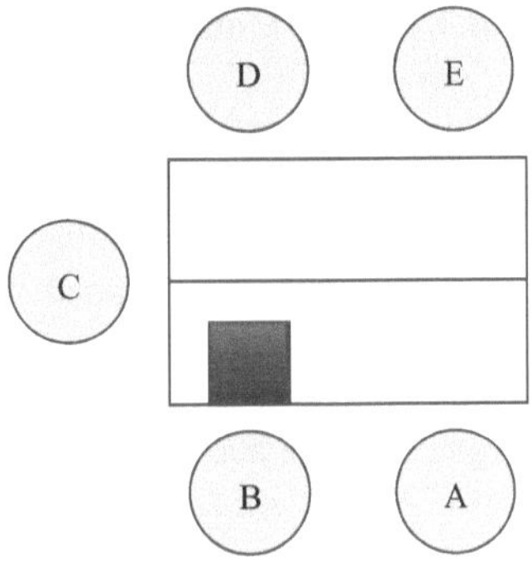

the rest of the lab. The lab is more permeable, as the students can easily look out the windows without being noticed. Students D and E have a position which allows them to oversee everything that happens within the boundaries of the room, as well as some of the outside. During the introduction of the material artefacts needed for the electrolysis activity, the teacher moved to the back of the room, standing almost next to this group, and students D and E needed to turn around on their swivel chairs.

The most prominent elements in this situation are the material artefacts used to show electrolysis – thereby creating a hub of activity. The students lean over the artefacts. They look at, smell and talk about the artefacts. Everyone except Student E. This creates a centre of practical activity near Student B. Student B controls the practical part of the activity and is mostly engaged with handling the material artefacts, spending much time in close observation (looking down into the beaker). Student E, sitting diagonally across, would have to raise or sit on the table to interact with the material artefacts used in the activity. Student E does not speak "subject matter" during this activity. Does the physical distance make it harder to get involved in the science practice? In many respects, the distance across the table is a short one. However, as the artefacts are small, the relative distance becomes larger. It becomes a question of whether the physical distance too easily becomes a mental distance. For a student in Student E's physical position, it will be difficult to see and possibly also to smell – access is difficult. However, this can also be a great position if Student E wishes to relinquish control over the practice.

Student C and especially Student A do most of the disciplinary talking in this group. Their contributions often build on each other's and they generate explanations together. There is a sort of axis between these two students, as they seem to share eye contact with each other to a greater extent than, for instance, with Student E or Student B. Student D asks questions and provide some comments, as does Student B. Student A uses a computer to write down results and take notes. None of the other students can see what is written on the screen. The screen acts as a sort of theoretical centre in this group since it functions as a source for information. While this student occasionally involves the other students in oral construction of descriptions and explanations, she/he creates a rather exclusive theoretical centre that the

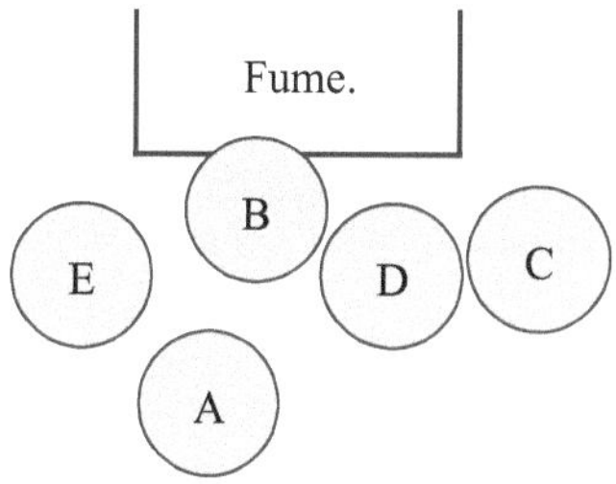

Fig. 5.3 Students placed around the fume cupboard

others cannot easily access. There are several other computers on their shared desk, though this is the only computer in use during the experiment. This creates a sort of centre for theoretical knowledge around Student A. This student becomes the one who initiates and sustains many of the conversations on knowledge, keeps the records and prepares the description and explanation to the report. When students have no access to, for instance, the written results, because they are on another student's computer – what then? It can be argued that this physical distance in combination with a student's location and direction makes the other students more disadvantaged in the production of their joint report.

5.4.3 Activity Location: By the Fume Cupboard

A while later Student B says: "Shall we just start then" and leads the way across the room to the fume cupboard on the other side of the science lab. Students C, D and E follow, while Student A writes down notes on the computer. Some of the students have "errands" on the way to the cupboard. These "errands" take them past other groups – but these are short detours. Student B arrives first at the fume cupboard and takes a clip and metallic magnesium. Students C and E place themselves on either side of Student B; see Fig. 5.3. Students D and E hold their phones ready to take pictures or videos of the burning magnesium. However, when it starts to burn, Student C and Student D have turned away – chatting. After one completed burning, Student A joins the rest of the group and stands behind the others, using the spaces in-between to have a peek at what is going on. Student C says "Since you weren't here, we'll do it once more". Student B burns one more strip. It flares up.

Table 5.2 and Fig. 5.3, respectively, show what the students are doing and where they are standing in front of the fume cupboard.

Student A is not handling any of the artefacts and is merely a spectator at this activity. The other students are engaged in different disciplinary actions using the artefacts they have available. However, Student B is the centre of activity. The walls and the shield of the fume cupboard create firm boundaries – although the shield is permeable visually, the access to the cupboard's interior is limited. It becomes quite crowded around the fume cupboard as indicated by Fig. 5.3. The students are crammed together and they divide tasks implicitly; see Table 5.2. There are short

Table 5.2 Students' disciplinary actions at the fume cupboard and their use of material artefacts for this purpose

Actor	Actions (disciplinary)	Material artefacts
Student A	Arrives late – watches	
Student B	Holds the magnesium	Magnesium and clip
Student C	Takes photos	Phone
Student D	Takes photos	Phone
Student E	Ignites the Bunsen burner	Match

distances. The student holding the magnesium has to be in front of the cupboard and not be squeezed by the others in accordance with health and safety precautions. In other words, Student B requires the most elbowroom; see Fig. 5.3. Student E is standing in front of the neighbouring fume cupboard, which is not in use at the moment. If it was, the student would have to stand closer to the others. As the figure shows, Student A and Student C cannot partake in the practical part of the experiment – they have, however, at least some possibility to gain access to what is going on by observing and recording.

Student B is dealing with the equipment. By leaving the desks first and arriving at the fume cupboard first, Student B can ensure a central position in carrying out this activity as well. Student B becomes the "doer". The centre of the activity is the material artefacts – controlled by Student B. All the other students are in a relative distance to this activity centre so their contributions can be verbal or recording the event. However, this is an activity that generates little talk. On the other hand, it becomes easy to engage more physically as it is easy to move around. Their positions shift slightly so they are able to see more, make better recordings or help ignite the metal. However, it is difficult to gain access to the artefacts if a student is standing in the back row. The cupboard can thus act as a maker of disciplinary disadvantages. For the students this might of course be seen as an advantage – not having to deal with science (artefacts).

During practical activities students have far greater potential mobility without being sanctioned. There is less regulation of students' bodies, and these students use this freedom in a "responsible" manner. They do not make long detours, nor do they "run around". There is some small talk with other students in the vicinity, but mostly they face the task they are working with.

5.5 Hybrid Space and the Production of School Science

In the previous sections, I have argued by use of literature as well as empirical observations for the particular hybridity of the school science lab and how this physical space is remade by teacher and students to be fit for science practices. They move furniture and the practices are located in different parts of the room. Some parts of the space have greater displays of "scientificness" than others. The space shifts between "school" and "science". Moreover, I have explored the science practices within this hybrid space by focusing on two areas: desk and fume cupboard.

The physical layout of these areas influences the distribution of advantages and disadvantages among the students and the students' possibilities for participating in school science practices. This creates a backdrop for the forthcoming discussion on what this might mean for the production of school science.

Leander et al. (2010) ask how to move beyond everyday notions of spaces of learning emphasising surface features such as classroom décor or human-centred theories of learning that reduce learning to a matter of relationships. While both décor and relationships between humans are important, they are not enough. Spaces of learning, in this case school science labs, provide certain possibilities for science practices and thus production of the school science subject. In his discussion of curriculum and research on mathematics education, Thomas Popkewitz (2004, p. 25) claims that "Contemporary pedagogical research tends to separate the issues of "knowledge content" from what a teacher [and students, sic!] does with that "content"". Although this claim is more than a decade old, it still holds relevance, also for school science. However, I claim that notions of content are also uncoupled from the spaces where they are dealt with.

By creating centres for theoretical and practical activity that are located at a distance from each other, there is a spatial separation of "practical" and "theoretical" learning. The theoretical belongs in front – the teacher's area (see also Hipkiss (2014) for similar findings). In the group work, the theoretical part of the practice is only partly distributed, as one student accesses the computer and theoretical information flow. When the students are working with the fume cupboard, they only "do"; they conduct an everyday observation in addition to recording the event, they do not write or make attempts on explanations. The explanations are left to the desk area. Thus, there are separated areas for meaning-making and practical activities. In other words, school science is produced through a separation of "theory" and "practical activities". Perhaps it would be easier to connect theory and practice for students if these two areas were collocated. I will argue that the separation of spatial practice is a direct consequence of the hybrid space. The double function provides some disadvantages for the science practice. For example, since this is a classroom there are school desks – and chairs. Students join together desks to sit at while they conduct their inquiries. If they are to look at each other – which is "normal" social behaviour, the rest of the group has no access to information on the group's computer. If everyone writes, it would be inexpedient. In a professional lab, this "problem" would probably be non-existent as there would be benches for the students to stand at as well as a clear division of labour. Moreover, they could have moved more easily to focus on the artefacts and computer. However, the main role of a classroom is to facilitate learning, and by letting the students work in quite large groups and look at each other, it is possible to facilitate a physical environment where different views are shared and explicated; see also Johansen, Jónsdóttir and Kolstø (this volume).

School has a regulating function (Nespor 1997). In other words the production of school science has to relate to the socialisation of students – regulating their bodies – and not only science content learning. The hybrid space's permeability plays a part in the regulation and control of bodies. The students are always on display, as people on the outside might watching. This is a particular kind of implicit social control.

Moreover, the permeability is both inside and at the intersection of inside-outside. Inside the school science lab, it is permeable in a different way than when it doubles as a usual classroom. While students are doing practical work, they can move quite freely, and they do. This provides a "freedom" that has to be connected to the production of science. Doing science creates opportunities to move. However, the teacher could have regulated the students in such a way that movement would be "unnecessary". The permeability inside the room – the possibility of moving around – makes new centres possible and creates new distributions of advantages and disadvantages.

Seen from a relational perspective, the division of labour within the groups can be interpreted as a sign of an individual learning paradigm colliding with a collaborative. Perhaps this is not quite so simple, perhaps built into the social space, there are features that support the individual – at the expense of collective meaning-making, e.g. computer screens and other smaller material artefacts. If the point of practical work is that all students can hear, touch, smell and see – the number of students in the group needs to be reduced – and they need to be physically close as well as be able to move about while interacting with the artefacts. However, if the groups are small, it becomes difficult to bring different voices into the discussion. Perhaps the lab itself is constructed to support the work of individuals or pairs and thus hinders the objective of "different voices".

The physical space plays a role in the production of school science. It projects a view of what science is – at least the school version of science. In this particular version, there was no place for physics or geoscience. In addition, a science lab where it is not possible to create a blackout is ill-suited to many different experiments. The light and windows are more "school" than "science", so to speak. One might wonder if the school practice of associating general science with biology creates a greater separation between students and disciplines such as physics, chemistry and geoscience. The space and its décor facilitates some styles of reasoning – of what and how science is seen.

The invisibility of the school science lab produces an influence on the school subject that is well worth further scrutiny. As long as the physical space is naturalised and there is a collective perception that influences the relation between the social and spatial dimensions, it also renders the production of school science partly invisible.

Acknowledgements There are two persons in addition to the researchers within the network that have made this text possible: Professor Stein Dankert Kolstø for sharing this interesting empirical material with me and Associate Professor Kristin Solli for valuable input and constructive criticism.

References

Abrahams, I. (2009). Does practical work really motivate? A study of the affective value of practical work in secondary school science. *International Journal of Science Education, 31*(17), 2335–2353.

Andersson, E., & Sørvik, G. O. (2013). Reality lost? Re-use of qualitative data in classroom video studies. *Forum Qualitative Sozialforschung/Forum: Qualitative Social Research, 14*(3).

Baird, D. (2004). *Thing knowledge. A philosophy of scientific instruments*. Los Angeles: University of California Press.
Carlone, H. B. (2004). The cultural production of science in reform-based physics: Girls' access, participation, and resistance. *Journal of Research in Science Teaching, 41*(4), 392–414.
Erickson, F. (2006). Definition and analysis of data from videotape: Some research procedures and their rationales. In J. L. Green, G. Camilli, & P. B. Elmore (Eds.), *Handbook of complementary methods in education research* (pp. 177–192). Mahwah: Laurence Erlbaum.
Farahani, F. (2010). On being an insider and/or an outsider: A diasporic researcher's catch-22. In L. Naidoo (Ed.), *Education without borders* (pp. 113–130). New York: Nova Science Publishers.
Gyllenpalm, J., & Wickman, P.-O. (2011). "Experiments" and the inquiry emphasis conflation in science teacher education. *Science Education, 95*(5), 908–926.
Hammersley, M., & Atkinson, P. (2007). *Ethnography: Principles in practice*. London: Routledge.
Hipkiss, A. M. (2014). *Klassrummets semiotiska resurser: en språkdidaktisk studie av skolämnena hem-och konsumentkunskap, biologi och kemi [Semiotic resources in the classroom: A language didactical study of the school subjects home economics, biology and chemistry].* (PhD), University of Umeå, Umeå.
Hofstein, A., & Kind, P. M. (2011). Learning in and from science laboratories. In B. J. Fraser, K. Tobin, & C. J. McRobbie (Eds.), *Second international handbook of science education* (pp. 189–207). Dordrecht: Springer.
Jobér, A. (2012). *Social class in science class*. (PhD), Malmö University, Malmö.
Jones, P. (2008). The interplay of discourse, place and space in pedagogic relations. In L. Unsworth (Ed.), *Multimodal semiotics: Functional analysis in context of education*. New York: Continuum.
Kind, P. M. (2003). Praktisk arbeid og naturfagvitenskapelig allmenndannelse [Practical work and scientific Bildung]. In B. Bungum, D. Jorde, & S. Sjøberg (Eds.), *Naturfagdidaktikk: Perspektiver, forskning, utvikling [Science education: Perspectives, research, development]* (pp. 226–244). Oslo: Gyldendal akademisk.
Knorr Cetina, K. (1999). *Epistemic cultures. How the sciences make knowledge*. Cambridge, MA: Harvard University Press.
Kress, G., & van Leeuwen, T. (2006). *Reading images. The grammar of visual design*. Oxon: Routledge.
Kress, G., Jewitt, C., Bourne, J., Franks, A., Hardcastle, J., Jones, K., & Reid, E. (2005). *English in urban classrooms. A multimodal perspective on teaching and learning*. Oxon: RoutledgeFalmer.
Latour, B., & Woolgar, S. (1986). *Laboratory life. The construction of scientific facts*. Princeton: Princeton University Press.
Leander, K. M., Phillips, N. C., & Taylor, K. H. (2010). The changing social spaces of learning: Mapping new mobilities. *Review of Research in Education, 34*(1), 329–394.
Lefebvre, H. (1991/1974). *The production of space*. Oxford: Blackwell Publishing.
Leijon, M. (2016). Space as designs for and in learning: Investigating the interplay between space, interaction and learning sequences in higher education. *Visual Communication, 15*(1), 93–124.
Nespor, J. (1997). *Tangled up in school: Politics, space, bodies, and signs in the educational process*. Oxon: Routledge.
Popkewitz, T. (2004). The alchemy of the mathematics curriculum: Inscriptions and the fabrication of the child. *American Educational Research Journal, 41*(1), 3–34.
Ravelli, L. J. (2008). Analysing space: Adapting and extending multimodal frameworks. In L. Unsworth (Ed.), *Multimodal semiotics: Functional analysis in context of education* (pp. 17–33). New York: Continuum.
Snell, J. (2011). Interrogating video data: Systematic quantitative analysis versus micro-ethnographic analysis. *International Journal of Social Research Methodology, 14*(3), 253–258.
Soja, E. W. (2010). *Seeking spatial justice*. Minneapolis: University of Minnesota Press.
Tiberghien, A., Veillard, L., Le Maréchal, J.-F., Buty, C., & Millar, R. (2001). An analysis of lab work tasks used in science teaching at upper secondary school and university levels in several European countries. *Science Education, 85*(5), 483–508.

Chapter 6
The Inquiry of the Cyclops: Dewey's Scientific Inquiry Revisited

Lars Bang

6.1 A Brief Diagnosis of Scientific Inquiry

This chapter aims to trouble the contemporary conceptualisation of scientific inquiry in science education and outline a direction for an alternative conceptualisation. I am mainly interested in the conceptual confusion regarding scientific inquiry as it is actualised in science education (what it is in relation to scientific method, scientific literacy, etc.) and the specific new emphasis on the development of mental abilities of the learner. This tension is highlighted below by Norman G. Lederman, Allison Antink and Stephen Bartos:

> Scientific inquiry has always been ambiguous in its presentation within science education reforms. In particular, inquiry is perceived in three different ways. It can be viewed as a set of skills to be learned by students and combined in the performance of a scientific investigation. It can also be viewed as a cognitive outcome that students are to achieve. (…) The third use of 'inquiry' in reform documents relates strictly to pedagogy and further muddies the water. In particular, current wisdom advocates that students best learn science through an inquiry-oriented teaching approach. It is believed that students will best learn scientific concepts by doing science. In this sense, 'scientific inquiry' is viewed as a teaching approach used to communicate scientific knowledge to students (or allow students to construct their own knowledge) as opposed to an educational outcome that students are expected to learn about and learn how to do so. (Lederman et al. 2014, pp. 290–291)

This important statement exposes the issues regarding contemporary scientific inquiry, especially when we consider John Dewey (1938) and Joseph Schwab's (1960) original intentions. The understanding of scientific inquiry appears 'stuck' and fluctuating between a method, a set of skills to be learned, a cognitive outcome and finally a specific teaching approach and pedagogy. In other words, scientific inquiry has become synonymous with general scientific activity and has lost its

L. Bang (✉)
Manchester Metropolitan University, Manchester, UK
e-mail: L.Jensen@mmc.ac.uk

K. Otrel-Cass et al. (eds.), *Cultural, Social, and Political Perspectives in Science Education*, Cultural Studies of Science Education 15,
DOI 10.1007/978-3-319-61191-4_6

specific distinctness, something other than method or methodology, as emphasised in the quote above.

My focus is on the general conceptual understanding and expression of scientific inquiry in science education, without considering in detail the various approaches to scientific inquiry or inquiry-based science education (see, for instance, Daphne Minner, Abigail Levy and Jeanne Century's overview of inquiry-based science instruction from 1984–2002) highlighting the above tension regarding inquiry in science education (2010). This problem regarding scientific inquiry as fluctuating between a teaching method and an inquiry approach was similarly raised by Gyllenpalm et al. (2010), but here I try to historically diagnose the problem as something stemming from a misinterpretation of John Dewey's original intentions.

Whilst there are differences between how inquiry is actualised in science education in various schools and scientific inquiry in scientific institutions, my focus is here primarily on the conceptualisation of scientific inquiry in science education. This troubling of the conceptualisation of scientific inquiry is taken based on the writings of John Dewey (1938), Joseph Schwab (1960), Roger Bybee (2004) and Norman G. Lederman and others (Lederman et al. 2014), and I use their writings to exemplify issues regarding scientific inquiry. I consider a mainly American and Anglo-Saxon approach to scientific inquiry/enquiry but acknowledge that different countries and cultures might have different expressions of scientific inquiry. Scientific inquiry is here very much related to inquiry-based learning and inquiry-based science education, but in a general way, as the specific expressions of above approaches might be different from the texts drawn upon in the diagnosis below. Then, I will outline a new way of conceptualising scientific inquiry, through the works and ideas of the two philosophers Benedict de Spinoza and Gilles Deleuze. I will position scientific inquiry within a flow or continuum of inquiry, where scientific inquiry is a specific actualisation. Additionally, I will forward an outline of a reconceptualisation of scientific inquiry resting upon a parallel approach to the general human activity of mind and body.

Scientific inquiry has taken a central role in science education in the last six decades, especially since Joseph Schwab and others proposed scientific inquiry as the very core and frame of the scientific enterprise and method. For instance, Alfred Novak's article on scientific inquiry highlighted this new central approach in science and science education (1964) and used Schwab's writing as the frame. Schwab argued that science itself had changed and that teaching in science education should adapt accordingly, 'The formal reason for a change in present methods of teaching the sciences lies in the fact that science itself has changed. A new view concerning the nature of scientific inquiry now controls research' (Schwab 1958, p. 374). Schwab argued for a change in science education, to teach science as inquiry, and this meant three things: (1) science in itself is presented as inquiry, (2) the student undertakes inquiries as a way to learn the subject matter, and (3) finally there is a revision of the traditional use of the classroom and laboratory (Schwab 1958, p. 377).

Central to the notion of scientific inquiry proposed in the 1950s and 1960s is the inspiration from John Dewey's thinking in his work *Logic: The Theory of Inquiry*

(1938) and from Karl Popper's *The Logic of Scientific Discovery* (2002 – English translation of 1932 original). Schwab's new direction for scientific education as inquiry was greatly inspired by Dewey's thoughts, and although Popper played a great role in the development of the philosophy of science, the focus will be on John Dewey, as he had the most direct influence in the development of scientific inquiry in science education. Dewey's notion of inquiry will be further expanded in the section 'Dewey's Theory of Scientific Inquiry [The Patient Zero]: A Realignment', but what is crucial here, in this brief diagnosis of the contemporary concept of scientific inquiry, is his novel connection between logic and subject matter. Dewey saw scientific inquiry as something greater than mere methodology and method and posited a particular relation between logic, subject matter and inquiry. When subject matter, for instance, subject matter in science education, is subjected to inquiry under proper conditions, logical forms 'accrue' and arise out of such a process (Dewey 1938, p.101). Dewey means that scientific inquiry can never be conflated to a teaching method, but should be a distinctive process engaging the subject matter, ending in a 'warranted assertion' (Dewey 1938, p. 144). This combination of a rationalist and an empirical approach was Dewey's philosophy of science, placing inquiry central and as something distinct. These thoughts were repeated and refined by Schwab (1960), Novak (1964) and Paul DeHart Hurd (1958) and other science education researchers in the 1950s and 1960s. Wesley Salmon (2006) identified a particular deep-rooted conflict regarding the status of language and logic in scientific explanation and discovery at this time – the conflict between semantics and pragmatics. Dewey's concept of inquiry and general philosophy of science is positioned as a philosophical pragmatic approach, blurring the line between a practical and normative judgement and inquiry. In contemporary science education, the role of scientific inquiry has been further stressed, especially after the US National Science Education Standards report in 1996 that posited scientific inquiry as referring to the 'diverse ways in which scientists study the natural world' (NRC 1996, p.23), whilst simultaneously linking this scientific inquiry to the activities of the students and their learning. In the NRC document, a link between the students' inquiry and the study of scientists is enunciated, but without stipulating this relation. The notion of inquiry is placed in the centre of scientific understanding and development of scientific knowledge. The process of scientific inquiry is referred to as a 'multifaceted activity' (NRC 1996, p. 23), but it is mainly a learnt cognitive activity and skill. For instance, science is described in the English GCSE curriculum as a way to 'develop and learn to apply observational, modelling, enquiry and problem-solving skills, both in the laboratory and in other learning environments' (Education 2013, p. 3). Similarly, the references suggested for further reading in the US National Science Education Standards report (NRC 1996, pp. 24–25) confirm this cognitive approach to the activity of inquiry, with the oldest reference being to Jean Piaget's work *The Construction of Reality in the Child* (1954). In other words, in the National Science Education Standards report (1996) and in the English GSCE curriculum (Education 2013), there is a suggested cognitive legacy and emphasis on explanation of the phenomena and activity of scientific inquiry as a skill to be learned, applied and assessed.

This has been further defended, enunciated and put in relation to contemporary scientific inquiry by Bybee (2004). Bybee stresses three important things that define the new scientific inquiry promoted by him and the US NRC.

The perspective presented here has several defining characteristics. Inquiry in the science classroom should:

- Focus on the learning more than pedagogy.
- Focus on learning outcomes more than the experience of students.
- Focus on science and learning theory more than teaching methods (Bybee 2004, p. 10).

Focus is here placed upon scientific inquiry as the teaching strategy, as something greater than 'student-originated' or 'hands-on activity' (Bybee 2004, p. 10), and the learners' mental capabilities to reflect and deduce other cognitive activities are central. This emphasis obscures the processual aspects of scientific inquiry and in fact reduces its overarching role in the development of knowledge.

Another contemporary example of the emphasis on scientific inquiry in science education curricula is how the changes in NSES (National Science Education Standards) were meant to promote scientific inquiry in the USA (Council 1996). Noticeable here and related to the issues pointed out previously is the change from implementing inquiry as a set of processes to implementing inquiry as instructional strategies, abilities and ideas to be learned, an issue also stressed by (Lederman et al. 2014). This shift now placed inquiry as something within the student, moving away from Dewey's stance of scientific inquiry as a process.

The main problem with this contemporary approach is the subtle turn towards the individual, the scientific learner, who now should acquire the techniques and skills of scientific inquiry and develop appropriate mental and cognitive abilities of deduction and scientific reasoning. The combination of rationalism and empiricism, envisioned by Dewey, is missing, replaced by an emphasis on the development of cognitive abilities necessary to conduct scientific inquiry.

There is thus a shift in the conceptualisation of scientific inquiry towards a specific scientific learner, possessing particular mental skills. This 'inquiring' student becomes the new ideal and outcome in science education and is here be referred to a specific utopian template of being scientific or homo empiricus (Bang 2014). This new scientific learner can then be identified through international comparative tests, for instance, PISA, turning scientific inquiry into a tool of the state and the globalised economy, another outcome for the educational system in the higher education arms race. To resist such a development and potential use of scientific inquiry, I propose reconceptualising scientific inquiry, and in the next section, I will draw attention to Gilles Deleuze's concept of the dogmatic image of thought, as something that is intrinsic to the conceptualisations of scientific inquiry highlighted earlier.

6.2 The Dogmatic Image of Thought Related to Scientific Inquiry

A concept like scientific inquiry does not arise out of nothing, but is historically tied to the development of science and other types of related inquiry (philosophical inquiry, theological inquiry and so forth). Scientific inquiry is thus related to how thought in general has been conceptualised and linked to overarching development of human knowledge and civilisation.

In other words, the conceptualisation of scientific inquiry rests upon how thought is conceptualised both in science and in philosophy. Philosopher Gilles Deleuze (1994) argues for a specific dogmatic image of thought that haunts the history of philosophy, and the philosophy of science, and thus shapes the specific concept of scientific inquiry:

> According to this image, thought has an affinity with the true; it formally possesses the true and materially wants the true. It is in terms of this image that everybody knows and is presumed to know what it means to think. Thereafter it matters little whether philosophy begins with the object or the subject, with Being or with beings, as long as thought remains subject to this Image which already prejudges everything: the distribution of the object and the subject as well as that of Being and beings. We may call this image of thought a dogmatic, orthodox or moral image. It certainly has variant forms: 'rationalists' and 'empiricists' do not presume its construction in the same fashion. (Deleuze 1994, p. 131)

This dogmatic image of thought is something 'which already prejudges everything' (Deleuze 1994, p. 131) and blocks a new way of conceptualising thought. The dogmatic image of thought is based upon eight postulates that together form the overall dogmatic image of thought. In relation to scientific inquiry and how inquiry (or perhaps more adequately the inquiring mind) is framed in a particular image of thought, I would like to draw a particular attention to postulates five to eight in Deleuze's (1994) description:

> (5) The postulate of the negative, or of error (in which error expresses everything which can go wrong in thought, but only as the product of external mechanisms);
>
> (6) The postulate of logical function, or the proposition (designation is taken to be the locus of truth, sense being no more than the neutralised double or the infinite doubling of the proposition);
>
> (7) The postulate of modality, or solutions (problems being materially traced from propositions or, indeed, formally defined by the possibility of their being solved);
>
> (8) The postulate of the end, or result, the postulate of knowledge (the subordination of learning to knowledge, and of culture to method) (Deleuze 1994, p. 167)

I see these four aspects of the dogmatic image of thought, being related to inquiry, because they attach a specific kind of judgement derived from the questioning proposition to the process or activity of scientific inquiry. This is actualised in public examinations, where closed questions are posed to students, for instance, in the examinations for GSCE science (Education 2013) where students are asked to only give one reason why energy decreases along a food chain (AQA 2014). A negative judgement steers the inquiring mind or thought, sorting what is deemed scientific knowledge or not. This particular judgement is actualised in science education and scientific inquiry, limiting the general scope of the inquiry and turning inquiry into a method where the students are searching for specific, already known, answers and facts. Dewey tried to escape this problem regarding the logic of language, by pointing out logic and truth (or as he called it 'warranted assertion') accruing from subject matter and the inquiry process itself. But in the contemporary description of scientific inquiry, the dogmatic image of thought is now conceptualised with a new emphasised cognitive element, the development of the proper mental skills of the learner, as both the outcome and frame for scientific inquiry. To reconceptualise scientific inquiry, I mainly focus on the issues of judgement and inquiry to trouble scientific inquiry and reconnect it to a more general notion (continuum) of inquiry. To outline such a reconceptualisation, I propose here to see the notion of judgement in scientific inquiry related to the antique story of Odyssey and the cyclops, acting as both a vehicle to enunciate the inadequacies of the potential former dogmatic image of scientific inquiry and a crux to see scientific inquiry in a new conceptualisation.

6.3 The Inquiring 'Cyclops'

> 'Cyclops, you ask my name and I
> will tell it you; give me, therefore, the present you promised
> me; my name is Noman; this is what my father and mother and my friends have always called me'. (Odyssey)

The cyclops (Polyphemus) is not an arbitrary image postulated as the dogmatic image of thought of scientific inquiry. The cyclops, Polyphemus from Homer, is precisely the image of a savage questioning monster, half-blind and in the tale tricked by its inadequate questioning and judgement by the hero Odyssey. In other words, the cyclops as described by Homer is a monstrous image of a thought, an example of thought misled and inadequate to the civilised and rational Odyssey. In Homer's tale of Odyssey's encounter with the cyclops, Polyphemus, I wish to highlight four postulates regarding the dogmatic image of thought and how they are specifically related to scientific inquiry. First, we have *(5) the postulate of the negative* of the monstrous hungry cyclops, mistrusting Odyssey and his crew, eating them for any sign of slight or rebellion and trapping them in his own cave. Second, we have *(6) the postulate of logical function, or the proposition*, seen in Polyphemus' sensible question to Odyssey 'What is your name?', to which Odyssey answers 'My name is Noman'. The cyclops is here misled by his own inquiring question and proposition. Third, we have *(7) the postulate of modality, or solutions,* represented in the very composition of the monster; it is a giant with one eye. It has trouble seeing and uses its strength to dominate its surroundings, not its cunning, reason or *metis* like Odyssey (Kershaw 2007). Scientific inquiry has in the dogmatic image of thought become such a one-eyed monster, forcing everything to abide by the monster's logic, rules and solutions. Finally, we have *(8) the postulate of the end, or result, the postulate of knowledge*, encapsulated in the security of the power the cyclops thinks it holds over Odyssey, only by knowing his name, relying on the proposition and the limited gaze. The image of the cyclops enunciates the problems regarding scientific inquiry in the grotesque image of the monster, particularly the inadequate judgement connected to the cyclops' inquiry. Additionally, the tale of Odyssey and Polyphemus could potentially help reconceptualise scientific inquiry in science education towards an approach more centred around sensation, experience and experimentation rather than questions. To understand the need for such a direction of the reconceptualisation, scientific inquiry in science education is revisited through Spinoza and Deleuze.

6.4 Scientific Inquiry Revisited Through Spinoza and Deleuze

To conceptually reframe scientific inquiry, I turn to Gilles Deleuze and Benedict Spinoza. What is similar to such an approach is to see inquiry, as John Dewey did, as something distinctive and much more fundamental, and an absolute intrinsic part of existence. It shifts focus from the mental capabilities of the learner towards their sensation, experience and active experimentation. I want to reconceptualise inquiry in terms of flows and within a continuum of inquiry, where the actualised and contemporary scientific inquiry is a part of such continuum and flow, expressing a specific kind of inquiry, connected with the scientific episteme overall (scientific knowledge, method, rules and so forth). Inquiry conceptualised as a flow is inspired by Deleuze and Guattari's usage of flows (Deleuze and Guattari 1983) and specifically Deleuze's notion of experimentation derived from his reading of Spinoza (Deleuze 2001, p. 125), replacing discrete categories. Here, inquiry is similar to Deleuze's notion of experimentation, a combination of experience and experiment in the same concept, an approach both from the sensing/inquiring body and the inquiring mind. The Deleuze-Spinoza approach to inquiry is thus generally an expansion and reconnection of scientific inquiry to other forms of inquiry, coupling the rational nature of scientific inquiry with the compulsory start from bodily experience.

The four postulates highlighted above in the image of the cyclops are similarly overturned by the Deleuze-Spinoza approach. The postulate of the negative (5) in thought is reframed to be an absolute affirmation of thought. Thought can never be an 'error', but it can be inadequate in its understanding of nature, and imagination and passion (the human condition or existence for lack of a better word) are both the obstacle and the drive towards a more adequate understanding. When we understand rationally, our passions become active, and 'error' and the specific confused composition of the mind/body become precisely the frustration needed to reach a more adequate understanding. Thought is never seen as erroneous but as part of nature and the human condition, with its specific potentials and limitations.

The postulate of logical function, or the proposition (6), is reframed as a bracketing of language and the focus on learning (or testing) of formal elements of scientific inquiry in science education. Instead, the Deleuze-Spinoza reframing focuses on sensation, experimentation and experience as the basis of all inquiry, especially scientific inquiry. Sense is not to be found in language but in sensation, experimentation and experience. The affections of the body, our lived life in nature, are the premise from which all understanding begins. Science is not merely a language or a logical form to be learnt and memorised.

The postulate of modality, or solutions (7), is reframed, to seeing a multiplicity of approaches for a never-resolved problem and curiosity. This Deleuze-Spinoza-inspired shift changes the premise of seeing everything with one wounded eye, to seeing with faceted fly eyes, potentially escaping the human point of view. The conceptualisation of inquiry moves from the search for a formalised solution to open experimentation.

The postulate of the end, or result, the postulate of knowledge (8) is here reframed towards a certain kind of unfinished uncertainty connected to all forms of scientific inquiry. Nothing ever ceases and rests, especially thought and the inquiring mind. Every inquiry is based upon relational knowledge, and if one of those presumed scientific truths is found to be false, the whole relation must be readdressed and rethought. This is a shift from universal knowledge to specific knowledge and multiplicities, stressing the fallible point of view connected with the human existence and simultaneously the immense potential for understanding.

Understanding concepts in flows or degrees is a prevalent feature of such scientific inquiry. Inquiry as flow means students approach subject matter in degrees of inquiry, from posing questions to experiencing phenomena directly, to experimentation, etc. The concepts derived from such processes are similarly in a flow, meaning that understanding, e.g. the concept of waters' various states, is placed in a flow from seeing, swimming, heating, molecular structure, etc.

We thus never have a discrete activity of 'scientific inquiry', but instead an activity exists within a continuum or flow of inquiry. This new flow or continuum of inquiry is an ontological term, meaning that in Spinoza's terms inquiry is connected or, perhaps, *is* the activity of mind and body. Coupling inquiry with Spinoza's concept of activity and his ontological system allows inquiry to be seen as a parallel activity, always undertaken from two simultaneous activities:

1. An activity of the mind
2. An activity of the body

Inquiry is thus not only essentially human but also 'beyond' human, and one could reasonably argue that from such a perspective all living things have an activity of inquiry in the continuum. The specific human inquiry could potentially result in three kinds of knowledge, and I draw here upon Margaret Wilson's (1995) interpretation of Spinoza's epistemology, depending on the degree to which the relation and objects of inquiry are ordered in terms of the intellect (reason) and bodies (physical laws, affects and generally all spatial, extensive laws). The first degree of knowledge is where we rely on the mutilated and confused random sense perceptions of the body and signs/words. The first degree of knowledge is referred to by Spinoza as being 'opinion' or 'imagination' (Spinoza 1996, p. EIIP40S42). The second degree of knowledge is reason; this is where we form common notions and adequate ideas regarding the properties of things and form and see relations adequately. The third degree of knowledge is called 'scientia intuitiva' or intuitive knowledge, a kind of intuitive science, where you know the essence of things. What is specific regarding Spinoza's epistemology is the crucial nature of the dual ordering of things through reason and through affections of the body. This means scientific inquiry becomes a dual activity of the mind and body, connecting the activity of reason with bodily affects and experiences, both being necessary to gain knowledge and order the encountered phenomena into the order of the intellect and the order of bodies. This dual activity is an intrinsic component of reconceptualising scientific inquiry and abolishing the specific emphasis on the mental abilities of the learner. Learning to swim and experiencing the various states of water are crucial elements of my reconceptualisation of scientific inquiry, directly connecting bodily experiences with how the mind understands the concept and phenomena. The emphasis on the body or the 'feeling brain' as enunciated by Antonio Damasio (2003) is crucial in reconceptualising scientific inquiry. Knowledge is always active in Spinoza's view; there is no knowledge without an active 'doing'. Inquiry is not a process of judgement but of understanding, and understanding (or Spinoza's knowledge) is a continuum, where the highest understanding is the most connected one and everything is thus seen in terms of greater or lesser understanding. To summarise, scientific inquiry is a specific actualisation of a continuum of understanding (in Spinoza's view); it is the processual goal within a continuum or flow. Scientific inquiry is thus both an experiment and experience seen actualised in flows, drawing upon Deleuze's and Guattari's line of thought. The image of the cyclops focused on four postulates of thought specifically related to scientific inquiry; these postulates can be reframed in this Deleuze-Spinoza approach. In Dewey's *Logic: The Theory of Inquiry*, there is an intention of envisioning inquiry in such terms, and this will be unpacked below to enunciate the need to slightly correct Dewey's notion of 'matrix of inquiry' towards a Spinozist notion of understanding coupled with Deleuze's thinking in terms of flows. In other words, the argument here is to go back to Dewey in order to move forward with a more adequate understanding of scientific inquiry.

6.5 Dewey's Theory of Scientific Inquiry [The Patient Zero]: A Realignment

Revisiting Dewey's theory of inquiry is crucial for going forward with a reconceptualisation of scientific inquiry in science education. By readdressing and realigning the very foundation of scientific inquiry and how it is seen in science education, one can move forward. The realigning proposed is thus based upon the revisited scientific inquiry unpacked above, but going 'through' Dewey instead of ignoring his legacy and body of work. Such a strategy has been utilised and dubbed *affirmation* by Gilles Deleuze in his reading of Friedrich Nietzsche (Deleuze 2006), David Hume (Deleuze 1991), Immanuel Kant (Deleuze 1984) and others. Dewey is 'Patient Zero', because the issues he faced have not yet been adequately addressed and diagnosed. In other words, Dewey's scientific inquiry is both the cure and the disease (problem) of science education, and he is revisited and affirmed through a Deleuze-Spinoza approach. Although Dewey forwarded a new way of seeing scientific inquiry, it got caught up in the dogmatic image of the cyclops, reproducing an inadequate conceptualisation of inquiry.

To unpack and realign Dewey's conceptualisation of scientific inquiry, I turn to *Logic: The Theory of Inquiry* (1938) (henceforth *Logic*) and Matthew Brown's erudition of Dewey's inquiry (2012). Many of the perspectives in *Dewey's Logical Theory: New Studies and Interpretations* by Burke et al. (2002) have been helpful in my reading of Dewey's *Logic*.

Caveat emptor! My reading of Dewey's *Logic* is driven by a particular interest, of changing or overturning Dewey's concept of scientific inquiry with Spinoza and Deleuze sitting on my shoulders. It is thus a troubled reading, like the devil reads the Bible, to draw forth a Deleuze-Spinoza undercurrent in Dewey's work, to overturn his concept of scientific inquiry and to turn it away from the school of logic, to whom Dewey was addressing his writing. Dewey's article regarding Spinoza 'The Pantheism of Spinoza' is the rationale behind this specific overturning of Dewey, to readdress and reconnect his reading of Spinoza with his concept of scientific inquiry (Dewey 1882). I will focus on four key points from Dewey's *Logic*:

1. Inquiry as something different and above methods and methodology
2. Dewey's notion of a 'matrix of inquiry'
3. Dewey's relation between inquiry and logic
4. The end of inquiry in a specific type of judgement

These four points will be addressed to contribute to a reconstruction of Dewey's concept of scientific inquiry in science education. They are similarly all related to the four postulates of the dogmatic image of thought (the cyclops), given a particular expression in Dewey's *Logic*.

To assist the reading of the four key points of Dewey's inquiry, the following approximates the model Dewey presented in his *Logic* (Fig. 6.1) which is inspired by Brown (2012, p. 280).

In the above model, we can see an image of Dewey's process of scientific inquiry but generalisable to all forms of inquiry. The scientific inquiry process starts from an indeterminate situation and then proceeds in a sequence of inquiry (the linear stages), which again proceed in a nonlinear fashion, fluctuating between the various activities (Brown 2012).

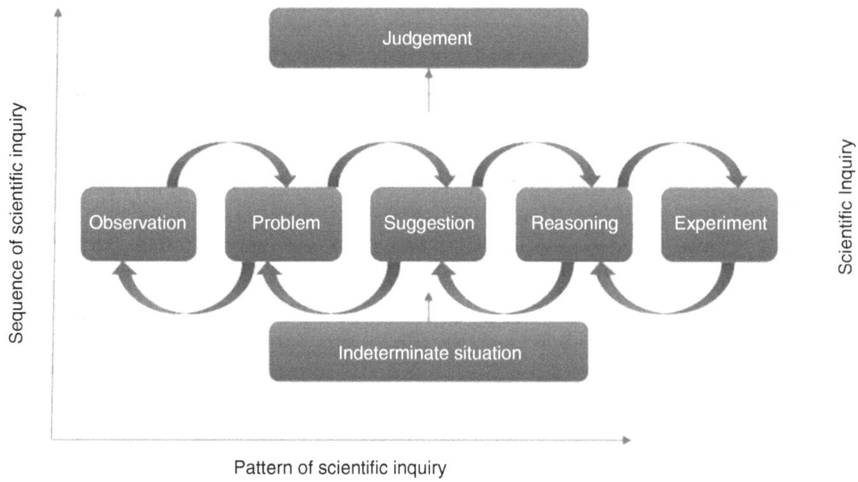

Fig. 6.1 Dewey's process of scientific inquiry

6.6 Inquiry as Something Different and Above Methods and Methodology

Dewey's logic is the essence of his earlier writings regarding his notion of inquiry, experience and philosophy of science. The *Logic* is Dewey's specific kind of inquiry, presented as a kind of amalgam between a logic and a theory of (scientific) inquiry. In other words, Dewey's logic, as it is presented in *Logic*, is his specific inquiry.

The aim of the *Logic* seems to be a dialogue and accounts between Dewey and his contemporaries in the philosophy of science. It is almost as if Dewey's writing is directly in dialogue with the American philosophical school of logic, particularly Rudolf Carnap's response to Dewey's *Logic* (1962), seems to support that reading.

> The theory, in summary form, is that all logical forms (with their characteristic properties) arise within the operation of inquiry so that it may yield warranted assertions.[…] To employ a convenient expression, it means that while inquiry into inquiry is the *causa cognoscendi* of logical forms, primary inquiry is itself *causa essendi* of the forms which inquiry into inquiry discloses. (Dewey 1938, pp. 3–4)

This quote is critical to understand Dewey's venture. It shows how closely Dewey's thinking is to a Spinozist notion of understanding, thus potentially acting as a vehicle and argument for bringing them more adequately together. We see that Dewey, akin to Spinoza, is fusing the process of methodology with the inherent logic discovered due to that process, meaning that methodology becomes intrinsic to the approach itself and not something 'applied' or 'operationalised'. Dewey's usage of the terms *causa cognoscendi* (the reason for knowing) and *causa essendi* (the reason being) is quite peculiar, and I see it as a reason to posit inquiry in slightly idealistic or metaphysical terms, inquiry as having a particular essence. My reading thus exposes the heritage of the early Dewey, who was much inspired by Hegel (Dewey 1930), and opens this quote for an encounter with Spinoza. In other words, Dewey's 'holistic approach' in regard to methodology and logic is seen here as a Spinozist approach, drawing upon a particular relation between the ontology and epistemology as something intrinsic to each other, not setting up a binary relation between ontology and epistemology.

For Dewey, the process of scientific inquiry never settles and only rests to be available as a point of departure for further inquiry, a 'warranted assertion'. Inquiry is thus separated from two seemingly obvious outcomes of inquiry belief and knowledge (Dewey 1938). Inquiry is not proposed as a method either, which again is similar to Spinoza's critique of method in *Treatise of the Emendation of the Intellect* (Curley 1985) and the rationalism inherent in his thought. Again, one can recognise Dewey's approach towards the role of method in relation to epistemology and ontology, resembling a Spinozist stance. To summarise the realignment here thus retains the notion of inquiry as something qualitatively different from a teaching method. To wrest Dewey's notion of inquiry from the cyclops' dogmatic image of thought, it is necessary to forward it, not as a specific logic but rather an inherent order and reason in the subject matter itself (nature) unpacked through a rational/bodily Spinozist inquiry. Such an approach is carried over into a Dewey's matrix of inquiry.

6.7 Dewey's Matrix of Inquiry

Dewey's matrix of inquiry is directly inspired by the philosopher and scientist Charles Peirce (1839–1914), stressed by Cheryl Misak (2004) and others, and contains an existential matrix of the biological and the cultural that together lead him to propose 'logic is the theory of experiential naturalistic subject-matter' (Dewey, 1938, p. 161). The matrix of inquiry has different areas, creating a type of 'situationism' or 'contextualism' associated with the inquiry process (Brown 2012). For Dewey, the biological and cultural areas of the matrix of inquiry are crucial, and he advocates for a naturalistic logic, connected to biological processes and development: 'The primary postulate of a naturalistic theory of logic is continuity of the lower (less complex) and the higher (more complex) activities and forms' (Dewey 1938, p. 23).

The cultural existential matrix of inquiry is situated in language, communication and the intellectual process. Dewey additionally assigns a specific importance to common sense as a distinctive area of the matrix of inquiry and a rudimentary form of inquiry. Overall, the matrix of inquiry results in a judgement. To wrest this part of Dewey's conceptualisation of inquiry away from the dogmatic image of the cyclops, especially regarding the postulate of modality, Dewey's matrix is slightly changed through Deleuze's notion of experimentation, stressing the inherently bodily experiences connected to inquiry. It is thus more a Deleuzian ethology (Deleuze 1988) than a contextualisation, highlighting the affective side of experiences with 'naturalist subject matter'. Such an inquiry is still open for connections between the biological and cultural areas, but using the bodily experiences as the starting point, not common sense connections stemming from language.

6.8 Dewey's Relation Between Logic and Inquiry

When you perform inquiry, logical forms attach themselves to subject matter, not a priori, but through the operation of controlled inquiry itself. The issue is that although Dewey fails to provide an adequate (meta)physical explanation or scientific explanation for that parallel occurrence, in many places Dewey hints towards a certain dialectic but never enunciates it fully (Brown 2012). The logical forms appear to emanate from nowhere, giving an unexplained ideal status to the process of inquiry itself.

In Dewey's argumentation regarding inquiry, he sees it crucial to distance himself from a purely rationalist perspective (that he judges Spinoza to be), and he keeps trying to balance his scientific inquiry between an empirical and rationalist stance (Dewey 1938, p. 10). Instead of advocating reason, Dewey posits the concept of inquiry as the overall denominator for a continuous process leading to 'warranted assertion' (Dewey 1938, p. 7). Dewey keeps holding out for his stance between the empirical and rationalist, regarding primary logical principles:

> Neither the existence nor the indispensability of primary logical principles is, then, denied. The question concerns their origin and use. In what is said upon this I follow in the main the account given by Peirce of 'guiding' or 'leading' principles. (Dewey 1938, p. 12)

Dewey's clearly stated that legacy to Peirce is highly relevant and interesting for the argument forwarded here regarding Dewey and Spinoza, especially the way Peirce saw Spinoza as a pragmatist and a crucial influence on his thinking, and I draw here upon Shannon Dea's (2008) analysis of Spinoza's influence of Pierce. Dewey begins his argument regarding inquiry similarly to a form used earlier (see, for instance, his argument for the concept of interest in Dewey (1913)), where he is advocating and outlining two historical kinds of inquiry:

(a) A common sense inquiry
(b) A scientific inquiry (Dewey 1938)

First, in the early Aristotelian (and Greek) logic, scientific inquiry was subject 'under' the inquiry of common sense. Dewey is arguing for a unification of the two types of inquiry, again similar to his unification of interest and effort (Dewey 1913). Dewey is pointing towards a seemingly unified pattern regarding inquiry, assembled in various matrices of inquiry (Dewey 1938, pp. 161–162). In other words, Dewey is trying to synthesise a form of formal logic connected with a specificity regarding empirical subject matter, or more crudely, attempting to fuse a rational frame with an empirical one. Dewey's main effort here, to fuse logic with the experiential process of inquiry, leads him to posit the following crucial definition of inquiry that becomes the historical premise, event and statement:

> Inquiry is the controlled or directed transformation of an indeterminate situation into one that is so determinate in its constituent distinctions and relations as to convert the elements of the original situation into a unified whole. (Dewey 1938, p. 167)

This point of Dewey, the relation between logic and inquiry, is 'simply' changed by referring to a Spinozist notion of absolute rationality. The dogmatic image of the cyclops of Dewey's scientific inquiry is turned towards an insistence of the absolute nature of thought, particularly seen in the postulate of the negative. In a monist apprehension of the universe (meaning that everything is one ordered expressive system), cosmology, thought or order/emergent complexity is quite absolute and not random. This means that the human inquiry has a potential for tapping into this order, because we are part of the order of the universe. Einstein forwarded this particular Spinozist stance (and saw it as being compatible with a scientific understanding (or notion of scientific inquiry) when he said, 'I believe in Spinoza's God, who reveals Himself in the lawful harmony of the universe, not in a God who concerns himself with the fate and the doings of mankind' (to Rabbi Herbert Goldstein, 1929). For Spinoza God is synonymous with nature; thus there is no need to argue for a particular logic, because logic is already inherent in nature for us to uncover.

6.9 The End of Inquiry in a Particular Type of Judgement

The end of the process of inquiry leads to a judgement or, what Dewey earlier described as, a 'warranted assertion'. Dewey seems to be caught up in logic itself, and whilst advocating a synthesis between common sense and scientific inquiry, he ends up establishing a tribunal of judgement in what Brown describes as a specific kind of 'situationalism' (Brown 2012, pp. 268–269), clearly in line with Cartesian thinking (see also Thomas Burke's analysis of that Cartesian element in Dewey (1994)). Dewey's particular judgement and its ambiguity have also been noted by Ernst Nagel (1950). Dewey is forced to accept a 'particular kind of judgement', going against his initial 'warranted assertion'. This inadequate component of judgement in Dewey's concept of scientific inquiry is juxtaposed with Spinoza's concept of understanding below, to reconfigure Dewey's concept of scientific inquiry and the seemingly Kantian component of imposed judgement on inquiry, which he directly takes from his reading of Kant (Dewey 1884). In other words, the dogmatic image of thought in Dewey's inquiry, specifically here related to the postulate of the end, is here wrested from a notion of judgement towards a Spinozist conceptualisation of understanding that resolves the issues regarding artificial ends of scientific inquiry.

6.10 Dewey's Judgement Versus Spinoza's Understanding

Dewey scholars, consistent with his own autobiographical sketches (Dewey 1930), propose a movement in his writings from an early period, where he was influenced by idealism (Hegel) to a later naturalistic pragmatist stance (Shook 2002; Wilkins 1956). We can see that Dewey has clearly engaged with Spinoza in his earlier writings (Dewey 1882), where his reading of Spinoza is filtered through Hegel's pantheistic readings. Dewey writes:

> We might have known, *a priori*, that such contradictions must occur in a pantheistic system like Spinoza's. It rests upon the basis that the only real knowledge is immediate knowledge. In this case the Absolute becomes mere Being, an Abstract Universal, possessed with no determinations whatever, for determinations are in this case, he cannot account for particular concrete objects. The two elements are necessarily irreconcilable from such a standpoint as Spinoza's regarding knowledge. (Dewey 1882, p. 257)

This critique against Spinoza's absolute as 'mere being' and Spinoza's inability to 'account for particular concrete objects' is similar to Hegel's (mis)reading of Spinoza, and I draw here upon Yitzhak Melamed's analysis of Hegel's reading of Spinoza (2010), and, unfortunately, for Dewey, it blinds him later in his *Logic* when he explains the origin of logical forms in inquiry. If he had Spinoza's notion of being and God (or nature), he would not have had to resort to logic as a magical occurrence in inquiry. I propose that Dewey's judgement be replaced with Spinoza's concept of understanding and continuum of knowledge, to be placed 'properly' in a frame of parallelism. Dewey cannot 'have his cake and eat it too' in regard to his relation between inquiry, logic and subject matter (nature). Nature has to be understood in terms of a certain kind of parallel rationalism (or, in modern terms perhaps in terms of an overarching emergent order as stressed by Michael Fischer (2005)). This systematic order and 'essence' is precisely what is uncovered in the process of inquiry in a continuum of knowledge, where we have a more or less adequate understanding of this order of nature. Precisely, because our minds and bodies themselves are part of this order or whole, we have the potential to understand and uncover the 'face of the universe'. Spinoza attributes power to thought and the activity of mind that explains, for example, why Albert Einstein's ideas and physics only decades later could be proved to be real and adequate. The whole movement between what we can reasonably think up or assume and later prove through observations and experiments is adequately explained by Spinoza's approach.

6.11 Dewey's Inquiry Revisited in the Cyclops

Dewey's concept of scientific inquiry was outlined to possess at least four key components:

1. Inquiry as something above and beyond method or methodology
2. Dewey's notion of a matrix of inquiry

3. Inquiry imposing a certain judgement
4. Finally, inquiry being connected to logic

These four points have been slightly revisited below to present a new outline and conceptualisation of scientific inquiry in science education. All these points will be explained through the earlier unpacked dogmatic image of thought and the cyclops and the movement towards a new image of thought, where inquiry is potentially connected more adequately to mind and body:

1. Scientific inquiry, inquiry into nature, is still a primordial and antique endeavour. Thus, retaining the component of the concept connects scientific inquiry with the very ontological nature of the universe, of God, etc. Scientific inquiry can still be seen as a monstrous cyclops, something for us to engage, overcome and potentially understand; if not vigilant we are easily misled by our bodies and inadequate perceptions. Scientific inquiry is an actualisation within a continuum of flows and a continuum of inquiry.
2. Scientific inquiry is still conceptualised within a matrix of various forms of activities and histories. Dewey proposed two points of existential matrices, a biological and cultural one, and similarly opened for other types of inquiries (social, etc.). I suggest here an attentiveness to many kinds of activities, both human and non-human. Spinoza's notion of activity is precisely one that emphasises an overall activity in all things (to a greater or lesser degree).
3. The false judgement by the cyclops, misled by the naming of things and by a misperception through seeing only with one eye, is inadequate for understanding the activity and endeavour of scientific inquiry. So, in a way it was good that Polyphemus was blinded by Odyssey – when we relate it to overturning the dogmatic image of thought related to scientific inquiry. Because in that 'blinding' lies the very lever where we can reaffirm the concept of scientific inquiry and reconnect it with different components. I replace all forms of 'judgement' by understanding to lesser or greater degree, fertilising Spinoza's continuum of knowledge to Dewey's thought and reconnecting scientific inquiry to a parallel approach between mind and body and between questioning and sensing/experiencing. Only when desire is emphasised over reason can reason truly be set free.
4. Dewey's attempt to connect and articulate (scientific) inquiry as a form of logic is thus both subverted and reaffirmed. As Dewey proposed, and rightly saw, there is a logic to be discovered in the process of inquiry of subject matter, but the discovered logic is not the 'things of logic' but rather the amazing Spinozist parallelism between the ontological and relational logic of things and the logic of thought.

6.12 Science Education and the New Revisited Scientific Inquiry

This reconceptualisation of scientific inquiry will not revolutionise science education in any way, but rather emphasise a new approach and how a reconceptualisation subtly changes practices involving scientific inquiry. In the following, I will focus on two critical points and examples of how such practices could change like the cyclops. These two points are directly connected both to the confusion about what 'scientific inquiry is' as pointed out by Lederman et al. (2014) and to the changed emphasis towards scientific inquiry, promoted by NSES, as something other than a process.

Not 'what is scientific inquiry' but 'what can scientific inquiry do' in a classroom. Scientific inquiry in the classroom is never isolated but exactly as Dewey emphasised coupled with the child's (or student's) earlier experiences or inquiries. A concept is developed in a continuum, from early experiences to more developed concepts related to other concepts. Inquiry is similarly seen as a way of asking questions and asking those questions differently, and one of these new actualised inquiries is scientific inquiry connected to the episteme of science. Scientific inquiry is similar to concepts in a flow and connected to the other forms of inquiry, making the inquiries of the sensing body/perceptions a real part of the process of scientific inquiry. Retaining the joy of understanding of discovery is only done through linking and revolution of earlier experiences.

Scientific inquiry is always 'between' bodies and minds. This revisited scientific inquiry redirects the focus from the mental capabilities of the learner or student as something to be inquired doing scientific inquiry, to an emphasis of the phenomena themselves and how they connect with us more wholly through our perceptions and initial understandings. The body is the starting point of all inquiries including the scientific ones, emphasising a much more experimental and experiential approach throughout the curriculum, for example, focusing on the body's innate understanding of surface (walking), gestures and tool use (hand-eye coordination, perception) and other such linked phenomena.

References

AQA. (2014) General certificate of higher education higher tier A 2. AQA Retrieved from http://filestore.aqa.org.uk/subjects/AQA-SCA2HP-QP-JUN14.PDF

Bang, L. (2014). Welcome to school – welcome to the empire-building business: An exploration and expansion of Bourdieu's notion of field. *Waikato Journal of Education, 19*(1), 51–62.

Brown, M. J. (2012). John Dewey's logic of science. *HOPOS: The Journal of the International Society for the History of Philosophy of Science, 2*(2), 258–306.

Burke, T. F. (1994). *Dewey's new logic*. Chicago: University of Chicago Press.

Burke, T. F., Hester, M. D., & Talisse, R. B. (Eds.). (2002). *Dewey's logical theory: New studies and interpretations*. Nashville: Vanderbilt University Press.

Bybee, R. (2004). Scientific inquiry and science teaching. In *Scientific inquiry and nature of science* (pp. 1–14). New York: Springer.
Carnap, R. (1962). *Logical foundations of probability* (2nd ed.). Chicago: University of Chicago Press.
Council, N. R. (1996). *National science education standards*. Washington, DC: National Academy Press.
Curley, E. (Ed.). (1985). *The collected works of Spinoza. Volume 1* (Vol. 1). Princeton: Princeton University Press.
Damasio, A. (2003). *Looking for Spinoza*. New York: Harcourt.
Dea, S. (2008). Firstness, evolution and the absolute in Peirce's Spinoza. *Transactions of the Charles S. Peirce Society: A Quarterly Journal in American Philosophy, 44*(4), 603–628.
Deleuze, G. (1984). *Kant's critical philosophy*. London: The Athlone Press Ltd..
Deleuze, G. (1988). *Spinoza: Practical philosophy* (trans: Hurley, R). San Francisco: City Lights Books.
Deleuze, G. (1991). *Empiricism and subjectivity: An essay on Hume's theory of human nature* (C. V. Boundas, Trans.). New York: Columbia University Press.
Deleuze, G. (1994). *Difference and repetition* (P. PattonTrans.). London: Continuum International Publishing Group.
Deleuze, G. (2001). *Spinoza: Practical philosophy*. San Francisco: City Lights Books.
Deleuze, G. (2006). *Nietzsche and philosophy*. New York: Columbia University Press.
Deleuze, G., & Guattari, F. (1983). *Anti-Oedipus : Capitalism and schizophrenia*. Minneapolis: University of Minnesota Press.
Dewey, J. (1882). The pantheism of Spinoza. *The Journal of Speculative Philosophy, 16*(3), 249–257.
Dewey, J. (1884). Kant and philosophic method. *The Journal of Speculative Philosophy, 18*(2), 162–174.
Dewey, J. (1913). *Interest and effort in education*. London: Leopold Classic Library.
Dewey, J. (1930). From absolutism to experimentalism. In G. P. Adams & W. P. Montague (Eds.), *Contemporary American philosophy: Personal statements*. New York: Macmillian.
Dewey, J. (1938). *Logic: The theory of inquiry*. New York: Henry Holt and Company.
Education, D. O. (2013). *Science: GSCE subject content and assesment objectives*. London: Crown Retrieved from https://www.gov.uk/government/uploads/system/uploads/attachment_data/file/211219/GCSE_Sciences_final_updated.pdf
Fischer, M. D. (2005). Culture and indigenous knowledge systems: Emergent order and the internal regulation of shared symbolic systems. *Cybernetics and systems: an International Journal, 36*(8), 735–752.
Gyllenpalm, J., Wickman, P. O., & Holmgren, S. O. (2010). Teachers' language on scientific inquiry: Methods of teaching or methods of inquiry? *International Journal of Science Education, 32*(9), 1151–1172.
Hurd, P. D. (1958). Science literacy: Its meaning for American schools. *Educational Leadership, 16*(1), 13–16.
Kershaw, S. P. (2007). *A brief guide to the Greek myths*. London: Robinson.
Lederman, N. G., Antink, A., & Bartos, S. (2014). Nature of science, scientific inquiry, and socio-scientific issues arising from genetics: A pathway to developing a scientifically literate citizenry. *Science & Education, 23*(2), 285–302.
Melamed, Y. Y. (2010). Acosmism or weak individuals?: Hegel, Spinoza, and the reality of the finite. *Journal of the History of Philosophy, 48*(1), 77–92.
Minner, D. D., Levy, A. J., & Century, J. (2010). Inquiry-based science instruction—What is it and does it matter? Results from a research synthesis years 1984 to 2002. *Journal of Research in Science Teaching, 47*(4), 474–496.
Misak, C. (2004). *The Cambridge companion to Peirce*. Cambridge: Cambridge University Press.
Nagel, E. (1950). Dewey's theory of natural science. In S. Hook (Ed.), *John Dewey: Philosopher of science and freedom: A symposium* (pp. 231–248). New York: Dial Press.

Novak, A. (1964). Scientific inquiry. *Bioscience, 14*(10), 25–28.
NRC. (1996). *National science education standards*. Retrieved from Washington, DC.
Popper, K. (2002). *The logic of scientific discovery*. London: Routledge.
Salmon, W. (2006). *Four decades of scientific explanation*. Pittsburg: University of Pittsburg Press.
Schwab, J. J. (1958). The teaching of science as inquiry. *Bulletin of the Atomic Scientists, 14*(9), 374–379.
Schwab, J. J. (1960). Inquiry, the science teacher, and the educator. *The School Review, 68*(2), 176–195.
Shook, J. R. (2002). Dewey and Quine on the logic of what there is. In T. F. Burke, H. D. Micah, & R. B. Talisse (Eds.), *Dewey's logical theory: New studies and interpretations*. Nashville: Vanderbilt University Press.
Spinoza, B. D. (1996). *Ethics*. (E. Curley, Trans.). London: Penguin Group.
Wilkins, B. T. (1956). James, Dewey, and Hegelian idealism. *Journal of the History of Ideas, 17*(3), 332–346.
Wilson, M. D. (1995). Spinoza's theory of knowledge. In D. Garrett (Ed.), *Cambridge companion to Spinoza*. Cambridge: Cambridge University Press.

Chapter 7
In-Between Chapter: The Political in Science Education

Helen Hasslöf and Iann Lundegård

7.1 Why is the Political Perspective Important in Science Education?

Science and technology have been of crucial importance for the development of humanity. Throughout history, new knowledge and technological innovations have made it possible to raise the standard of living for new generations in different parts of the world. However, this has also led to an unfortunate acceleration in the use of the world's natural resources. Nevertheless, scientific and technological findings are often promoted as the drivers of development in our societal and economic systems.

As a result, science has also become an important part of standard education all over the world. Different scientific domains – developed throughout history – have been implemented and transformed into school subjects such as physics, chemistry and biology (Popkewitz 2004). In general, those are given privilege in education as subjects of high interest, enhanced by comparative international school science tests such as PISA and TIMSS (Biesta 2009).

Science has after all, not only brought the good to the world but also new challenges. The emergence of scientific and technological paradigms has implicated questions in relation to issues of sustainability that need to be problematized further (Latour 2008). For example, the development of gene- and nanotechnology has gen-

H. Hasslöf (✉)
Malmö University, Malmö, Sweden
e-mail: helen.hasslof@mah.se

I. Lundegård
Stockholm University, Stockholm, Sweden
e-mail: iann.lundegard@mnd.su.se

K. Otrel-Cass et al. (eds.), *Cultural, Social, and Political Perspectives in Science Education*, Cultural Studies of Science Education 15,
DOI 10.1007/978-3-319-61191-4_7

erated numerous ethical issues including decision-making on GMOs, cloning, and health issues. Likewise, our need for a sustainable energy supply has presented difficult choices such as nuclear fuel management and how to deal with climate change. In many places, widespread monocultures, soil-erosion and reduction of biodiversity are the other side of the coin when developing technology for agriculture, forestry, and aquaculture. These are just a few examples of the issues that modern societies need to deal with, and all are linked to discussions of sustainability and development. Each of these issues carries uncertainties, contradictions, and conflicting interests which need to be considered (Lundegård and Wickman 2007).

Aligned with the scientific revolution, there are also challenges for education as well as for educational research. Within every choice of knowledge-use and development there lies a decision that needs to be taken into consideration concerning its consequences. Since decisions related to the use of scientific innovations empower different progressions of economic, cultural, and social justice, they also need to be considered in ideological terms. The above-stated scientific breakthroughs are some examples where political dimensions are revealed through the embedded different and conflicting values and interests. Similarly, this means there are choices to be made in educational research associated with an ideological deliberation about what to expose.

Besides its position as promoting future societal improvement, science education has another important role to play. In decisions of everyday life as well as in relation to civil rights and to having an ability to position oneself politically, *scientific literacy* plays a vital part. Here scientific knowledge could be seen as a democratic privilege, which gives opportunities to citizens to critically consider the implications of scientific and technological development from different perspectives. In other words, it becomes important to reveal "the political" in science education and to be aware that scientific literacy is a form of political agency.

Societal change due to political encounters between people might stem from and develop through democratic processes. However, which policy should be conducted or how democracy should be designed is not given by nature (there is no political ontology or ontological policy). Instead, the policies and forms of democratic processes are in constant motion. "The political" arises where there is passion and where there is dissensus about how the passion should be addressed, or where to channel or direct protests against present forms of oppression (Mouffe 2000). Hence, each new generation recreates policy content as well as forms of democracy.

The political thus can be approached as a content as well as a condition. It could either focus on individual agency or on conflicting discourses, or highlight the encounter between the individual/s and the discourse/s. However, in what way *the political* has possibilities to emerge in science education and how this is approached in relation to students' agency as individuals in relation to existing discourses is an interesting question.

The following chapters address political perspectives in order to present a "troubling perspective" in an effort to reveal or unpack the (antagonistic) values embedded in science teaching. With a point of departure from these chapters we discuss different approaches of *the political* and problematize its meaning in relation to science education. Hence, the political perspectives are made visible through different perspectives and educational contexts and through the various theories and analytical tools which these chapters are based upon.

The first chapter: "*What's in it for me? – How does a professional development program (PDP) meet science teachers' carrier expectations*", aims to raise awareness of how we might value best-practice or success-stories from evaluations of bigger projects. Peer Daugbjerg and Martin Krabbe Sillasen question whether PDP's that are designed to accommodate and improve individual teachers' professional capacity actually improve teacher learning effectively. Often these types of PDP's prescribe what teachers should be like and how they can become these predefined professionals. Further, Daugjerg and Krabbe Sillasen put forward how typical PDP's also contain a discourse of idealised meaning making for the participating teachers. One of the main arguments in this chapter is based on an analysis that reveals how the evaluations of these programs have different results depending on which methodology and which perspectives are put forward in the evaluation process. In other words, when deciding on how to evaluate success and "effectiveness" of such programs, a crucial question is: from whose perspective? Daugbjerg and Krabbe Sillasen show how a balance between top-down (standard systemic assessment) and bottom-up processes (narrative inquiry) could give a more balanced picture of "success-outcome" in the evaluation processes of bigger projects.

In this chapter the political emerges as a critical perspective on how PDP's are used as political tools for implementing educational reform policy. For example, they comprise a top-down and generalizing perspective assuming that "one-size fits all", and the assessments of the program narrow the analysis of the learning-outcome. In our view, the political in this chapter works as an analytical lens to troubleshoot the existing perspectives, to show the different "views and truths" that are at stake, and how the more complex picture of different voices and experiences is excluded.

The question arises then, whether universal frames for such programs should be implemented and if so, what the purpose might be. Should education be promoted to socialize students into a given order, or should the students themselves be involved in challenging the universal regime in order to establish a more local and particular framework for goals and visions (c.f. Todd 2009)? We might question whether we should socialize our students to embrace the decisions prescribed by national and international decrees, or if science education, SSI, or ESD, should be seen through political lenses, as something that needs to be renegotiated by each generation. Should education aim to normativity, or should it embrace pluralism?

In the second chapter Gerd Johansen, Gudrun Jónsdóttir and Stein Dankert Kolstø partially challenge the position that socio scientific issues (SSI) are necessary in order to address the political in science education. Instead traditional science teaching is seen as an actual practice of citizenship. In their chapter "*Ordinary school science as an arena for enacting citizenship. Deliberative communication; possibilities and constraints*" they claim that traditional science teaching, despite its content, should be regarded as being actual practices of citizenship.

In the analyses, using Tomas Englund's (2006) framework, patterns of deliberation are analyzed whilst the students have to deal with disagreement, differing interests, and with making decisions in discussions where they cannot foresee the outcome and where they perhaps even might question the authority of teachers or subject matter itself. The communication process is interpreted in line with Englund's framework and described as an enactment of citizenship that prepares students for future participation in democracy. This is an interesting perspective of teaching situations, seen as enactment of citizenship. However, we try this argument by linking it to the educational philosopher John Dewey, who early on dealt with democracy in education (Dewey 1916). Dewey describes democracy as a basic "form of life" (p.80) in human interaction on public issues (Dewey 1954; Van Poeck and Vandenabeele 2011). Consequently, the impetus for a democracy is not the process itself, but rather the desire for justice and solidarity. In this view, questions concerning politics that require deliberation always start in common content matters, such as resource allocation, energy distribution, or sustainable development.

Issues of sustainable development in teaching are the focus of the last chapter, "*Political agendas and actors in science teaching*". Maria Andrée, Lena Hansson and Malin Ideland discuss political rationalities invited into classrooms through different learning materials including ecological footprint calculators. The analysis targets how a specific kind of citizen is "made up" through the calculations, and what political ideology such making of a sustainable citizen is influenced by. Through Michel Foucault's theories of governmentality (Foucault, Senellart and Davidson 2007) the authors analyse governing technologies used in the ecological footprint calculators to discuss how these operate through techniques that produce "truth" and thereby standards for the normal, and environmentally-friendly, "human being" (Hacking 2006). In this text the political perspective is shaped to discuss how the (sometimes hidden) political agendas in science education may be troubled, i.e. in analyses of structural power. This obviously points to the political in relation to education, and even to educational research. It demonstrates that educators might start to question more often whose interests are embraced through an application. In this example a lot of good effort towards making students aware of climate change is compromised because of the implicit normativity. In addition, another interesting discussion arises within this context concerning how the researcher's own use of a particular analytical framework might construe particular power relations and hence, how we as researchers also have to be responsible for the potential consequences of our findings and constructions of the "good guys" and the "bad guys" in the arena.

Finally, we would like to conclude this text with some thoughts about the political in relation to subjectification and identity. Considerations in this area raise questions about inclusion and exclusion. Hence, while identity creation is mainly based on already constructed identities and subjects, the subjectification puts focus on the intersections where new subjects have the opportunity to "become". These are processes we can only invite to take place (by addressing the students as political subjects) but cannot presuppose to happen.

In educational contexts (highlighted by socio-scientific issues) this issue is important because it concerns how the students are regarded within the learning situation: whether they are viewed as passive receivers of information, who later will be introduced into existing societal democratic decisions, or if the students from the beginning are considered as democratically legitimate individuals (Todd 2009).

One way to open up opportunities for *the political* to be given expression in science education is to create teaching situations that enable students' subjectification processes – i.e. to create moments where they are addressed and challenged as political subjects (Hasslöf and Malmberg 2015; Lundegård and Wickman 2012). Here the interest of *the political* takes departure from the reciprocal communicative action where the political identities become challenged and expressed. However, political perspectives may trouble science education in many different ways, as is reported in the following chapters. We are convinced that education ought to be a mode of providing critical intervention of what should be counted as vital knowledge into an overall context of diverse values and agencies.

As science teachers and researchers, we are challenged to scrutinize how democracy and *the political* is addressed in the classroom. Obviously, it is evident that there is a need to discuss how to create educational environments that not only allow, but also encourage students to criticize, renegotiate, deconstruct, and transform. Thus, the teacher's role is to direct situations and to prepare for encounters where antagonistic thoughts become expressed and listened to and dwelt upon and valued. But also, to identify how scientific knowledge might be(come) a tool for developing students' political awareness.

References

Biesta, G. (2009). Good education in an age of measurement: On the need to reconnect with the question of purpose in education. *Educational Assessment, Evaluation and Accountability, 21*(1), 33–46.

Dewey, J. (1916). *Democracy and education: An introduction to the philosophy of education*. New York: Macmillan.

Dewey, J. (1954). *The public and its problems* (Original work published 1927). New York: Holt.

Englund, T. (2006). Deliberative communication: A pragmatist proposal. *Journal of Curriculum Studies, 38*(5), 503–520.

Foucault, M., Senellart, M., & Davidson, A. L. (2007). *Security, territory, population*. Basingstoke: Palgrave Macmillan.

Hacking, I. (2006). Making up people. *London Review of Books, 28*(16), 23–26.

Hasslöf, H., & Malmberg, C. (2015). Critical thinking as room for subjectification in education for sustainable development. *Environmental Education Research, 21*(2), 239–255.

Latour, B. (2008). *What is the style of matters of concern? Two lectures in empirical philosophy*. Amsterdam: Van Gorcum.

Lundegård, I., & Wickman, P. O. (2007). Conflicts of interest: An indispensable element of education for sustainable development. *Environmental Education Research, 13*(1), 1–15.

Lundegård, I., & Wickman, P. O. (2012). It takes two to tango: Studying how students constitute political subjects in discourses on sustainable development. *Environmental Education Research, 18*(2), 153–169.
Mouffe, C. (2000). *Deliberative democracy or agonistic pluralism, Political science series 72*. Vienna: Institute for Advanced Studies.
Popkewitz, T. (2004). The alchemy of the mathematics curriculum: Inscriptions and the fabrication of the child. *American Educational Research Journal, 41*(1), 3–34.
Todd, S. (2009). Can there be pluralism without conflict? *Philosophy of Education Yearbook., 2009*, 51–59.
Van Poeck, K., & Vandenabeele, J. (2011). Learning from sustainable development: Education in the light of public issues. *Environmental Education Research, 18*(4), 541–552.

Chapter 8
Political Rationalities in Science Education: A Case Study of Teaching Materials Provided by External Actors

Maria Andrée, Lena Hansson, and Malin Ideland

8.1 Introduction

Many Western societies have a tendency to discuss school as failing to foster an interest in science as well as failing to produce employable, internationally competitive, emphatic and flexible citizens. School science is often discussed as outdated (detached from the modern world and modern work life), not interesting enough for young people and non-effective for the students' learning. This discourse of school failure has resulted in many initiatives at various levels of government and also opens up for external actors to contribute to science education.

Historically, external actors have always been involved in school; for example, religious organisations, local companies and non-governmental organisations (NGOs) have, in the past and in different ways, engaged in students' learning. However, drawing on Stephen Ball (2009), we claim that a shift in the responsibility of schooling is taking place. "Statework", in terms of educational governance, is now carried out through multiple actors and relationships – a public/private statework. The governing of education is thus distributed within a multi-layered network where the borders between the state and the market are wiped out. This invites new actors to help schools make sense of and manage educational policy and problems.

M. Andrée (✉)
Stockholm University, Stockholm, Sweden
e-mail: maria.andree@mnd.su.se

L. Hansson
Kristianstad University, Kristianstad, Sweden
e-mail: lena.hansson@hkr.se

M. Ideland
Malmö University, Malmö, Sweden
e-mail: malin.ideland@mah.se

K. Otrel-Cass et al. (eds.), *Cultural, Social, and Political Perspectives in Science Education*, Cultural Studies of Science Education 15,
DOI 10.1007/978-3-319-61191-4_8

Concerning science education, the involvement of external actors such as the scientific community (Andrée and Hansson 2014), NGOs (Ideland and Tröhler 2015) and industrial actors (Andrée and Hansson 2015) have been embraced rather uncritically. For example, in regard to the engagement of industrial actors, many initiatives including school programmes, theme days, visits to companies, competitions, festivals and teaching materials (cf. Teknikdelegationen 2010; Andrée and Hansson 2015) are commonplace. In addition, some NGOs have contributed with programmes for environmental and sustainability education (Goldman et al. 2013; Ideland and Tröhler 2015) and a curriculum initiative for introducing new pedagogies in technology education (Stables and Kimbell 2001). It has also been shown that various actors are engaged in complex networks where, for example, industrial actors, various levels of government, and academia are all engaged in joint initiatives (Andrée and Hansson 2015). This shows that even initiatives involving external actors rely to some extent on public funding.

The collaboration of schools with external actors such as the scientific community, industrial actors and NGOs has been emphasised on various policy levels – transnationally and nationally. With the UN programme, Agenda 21, collaborations between actors from education, the business sector and civil society are regarded as a precondition for sustainable development.

On the European level, the European commission launched the *inGenious* programme with the aim of focusing on "the contribution that the private sector can offer to science education at primary and secondary school level" (*inGenious* 2014). On the national level, Swedish schools and teachers are explicitly encouraged to cooperate with external actors. The Swedish national curriculum for compulsory school states that teachers are obliged to "assist in establishing contacts with … organisations, companies and others who can help enrich the school's activities and establish it in the surrounding society" (National Agency of Education 2011, p. 19). According to Stephen Ball (2009), the educational task is thus distributed into many areas, the state is recalibrated, and a network of actors are involved in the education of future citizens. For STEM education in particular, widespread engagement by various actors is justified by pointing out that these subjects are necessary in the knowledge economy (Ideland et al. 2016).

The discourse of helping schools solve the "science education problem" opens up for various economic and ideological interests to enter the classroom without being critically scrutinised, neither from the perspective of political agendas nor the potential forum for "advertising". Thus, the intention of this chapter is to contribute with an analysis of what political rationalities are invited into classrooms through these external initiatives and by what means. How do NGOs and industrial actors become unproblematic educational actors? What comes along with the use of materials produced by an organisation with an agenda external to education? In other words, our aim is to trouble the roles of external actors in school science. We do this by analysing two so-called ecological footprint calculators designed by external actors as resources for teaching about energy use and resources. One calculator is provided by the nonprofit environmental organisation, WWF, and the other is by the private energy company, E.ON. Departing from the analyses of the ecological footprint calculators, we aim to problematise the roles of external actors in school and how they become involved in the organising of school science. Furthermore, this chapter has a theoretical aim – to use and discuss tools for understanding more of the rationales for science education and how these open up and limit ways of thinking and being in the science classroom – in other words, how we can trouble the political rationalities in science education.

8.2 Theoretical Framework

To understand how the external actors become legitimised and viewed as natural in the formal school system and to learn how they influence science education today, we turn to Michel Foucault's (1983) theories on how power is exercised through constructing standards for what we call the responsible scientific citizen. Through defining normality and how the desirable citizen is shaped and fashioned, the possibilities for what one can say, do and even think are set up and limited. This is how we can be governed – through our so-called free will – which Foucault conceptualises as governmentality (Foucault et al. 2007).

Nikolas Rose and Peter Miller (2010) have outlined a framework for analysing governmentality at three levels – *political rationalities, political programmes* and *governing technologies.* A *political rationality* can be described in terms of an intellectual "apparatus for rendering reality thinkable" (Rose 1996, p. 42); it defines what is taken-for-granted, obvious and rational (Rose and Miller 2010). This rationality is characterised by and operates through a specific morality, epistemology and idiom (style of reasoning) which contribute to the construction of the rationality as obvious (Rose and Miller 2010). The second analytical level, *political programmes,* targets how the rationality becomes transformed into everyday life on a policy level. In education, the most obvious political programmes are curricula and syllabi. But here, in the domain of sustainability, documents like Agenda 21 and EU initiatives to invite different stakeholders into the educational system are also relevant. People's lives – in this case, in the context of science education – become possible to administer through programmes because they construct problems and solutions for society (Popkewitz 2008). Finally, and what is analytically focused on in this chapter, are *governing technologies* – technologies for the internalisation of the political rationality into individuals' "souls". The governing technologies, which can be subtle, define certain lifestyles, competencies, actions, emotions and personalities as normal and good, while others simultaneously become defined as abnormal and undesirable. Governing technologies like those used in the aforementioned ecological footprint calculators operate through techniques that produce so-called truth and thereby set the standards for what is considered a normal and environmentally friendly human being (Hacking 2006). The question is, what political rationalities are allowed to operate in school through the ecological footprint calculators?

Central to the most common political rationality of today – what is often called *neoliberal* or *advanced liberal* – is the trust in individual choices. Decision-making and responsibility has moved from the state arena to the individual. Rose (1999) expresses that one is obliged to choose the right life, and the framework of governmentality is useful for understanding the politics of the neoliberal society: how to make the individual will comply with the national will without any force (Ideland 2017; Popkewitz 2004). The ecological footprint calculators illustrate this neoliberal logic; through seeming objective and calculating, they speak with the voice of expertise. Via the calculators, the everyday habits of the individual are translated to reveal the exact amount of environmental degradation caused. According to Miller (2004), calculative practices should be analysed as "mechanisms through which programs of government are articulated and made operable" (p. 179). In general, the numbers that appear in the results are seen as objective and indisputable and function as a way of depoliticising the political through making them appear objective. Numbers appear to safeguard from political will and cultural differences (Rose 1991). This mathematical logic is also closely related to the technoscientific belief in objectivity and rationality (Latour 1999) frequently communicated in the teaching of science.

Thus, central to the notion of a calculator are numbers. Through their assumed objectivity, numbers function as what Rose (1999, p. 121) conceptualises as "centers of calculations" or places for producing standards for human beings with the result of producing self-calculating persons (Miller 2004, p. 180). This is done through a variety of economic metaphors. One such metaphor is what Peter Miller refers to as "accounting": the practice of keeping books to record income and expenses and to calculate practices in society. The bookkeeping makes things and acts visible, and consequently, also possible to govern. Another powerful economic metaphor is the notion of debt and the indebted human being (Lazzarato 2012). Italian sociologist Maurizio Lazzarato (2012) claims that at the core of neoliberal governance is the idea of debt: the indebted human being is also a governable, administrable person. This is in relation to the economy, but we claim that debt is also a frequently used metaphor for understanding issues of environment and sustainability. This means that environmental problems are conceptualised as a debt to the natural environment whereby economic terms such as "resources" and "consumption" are used. The language of economics carries a specific way of understanding the relation of humans to the planet. Finally, Rose (1991) discusses statistics and numbers as an ethical technology and rationalises an ethical approach to the world. Through the objectivity of statistics and numbers, a tool is provided for making ethical judgements which are detached from emotion, passion and non-reason (cf. Bennett 2001). In this chapter we reconnect the numbers to ethics and emotion.

8.3 Method and Outline

From these three technologies – accounting, debt and ethics – we intend to analyse what the centres of calculations do in terms of governing a specific kind of subjectivity: the disciplined, rational and responsible citizen. Through this seemingly rational way of handling environmental issues, external actors may find a way into education, especially given that the political has been hidden behind notions of objectivity. By scrutinising the governing technologies, we intend to discuss how it can be possible or even natural to use teaching materials produced by an organisation with an agenda of their own. Deconstructing this naturalness serves as a way to illuminate political rationalities governing schools through collaboration with external actors. Thus, in this chapter, we analyse two ecological footprint calculators: one produced by the Swedish energy company, E.ON, and one produced by the nonprofit organisation, World Wide Fund for Nature (WWF). Both E.ON and WWF have extensive educational programmes with an emphasis on sustainable development and science education. Education in sustainable development has become a focused area for WWF since the 1990s when, in parts of the UN report, Agenda 21, NGOs were pointed out as important actors in the efforts for a sustainable world. This is especially the case in the context of Sweden, where WWF Sweden calls itself one of the most prominent national WWF organisations when it comes to educational issues. On its corporate website, E.ON has a special portal for students and teachers where a range of teaching resources are provided. Here, they state, "We want to generate interest and awareness around energy. Getting young people to feel commitment to one of our greatest challenges – how we will sustainably supply the entire world's population with energy" (https://www.eon.se/samhaelle---utveckling/energikunskap/energikunskap-foer-skolor/utbildningsmaterial.html). For E.ON, the engagement in education may also be seen in light of what they formulate as the strive to "become the most liked partner for sustainable energy solutions (https://www.eon.se/om-e-on/om-foeretaget0/foeretagsfakta.html)". In particular, we focus on how the calculators function as governing technologies in relation to political rationalities.

We first analysed the calculators as if we were students working with the task of determining our ecological footprints. When responding to the questions in the calculators, we have tried to be as honest and informed as possible concerning our own consumption. As we explored the calculators further, we started to change details in our answers to figure out how the algorithm worked with the question of what answers made a difference to the result in the end. While calculating our footprints, the idiom of the footprint was analysed: How were the questions phrased? What types of words were used? How were the questions and the results illustrated? We have paid particular attention to the numbers and how the calculations were carried out and communicated. What roles did the numbers play? How did they operate as a governing technology? How did they use the economics language of accounting and debt? And how is a specific code of ethics – how to be and act as a responsible person – constructed through this language? In the analysis of the ethics produced, we also discussed our emotions while taking the tests and finding out the results: How did it feel to be pointed out as an environmental hero, or alternatively, a crook? How did the calculator target our emotions and with what consequences? Our emotions became a part of the analytical assemblage and acted as a kind of emotional data that we embraced rather than excluded (Eriksson 2016). To summarise, the analytical framework focused on what kind of political rationale – consisting of idiom (styles of reasoning), epistemology and morality – enters the science classroom through the calculators?

From this theoretical and methodological framework follows the result sections which describe and analyse the governing technologies of the ecological footprint calculators of "My climate footprint" by E.ON and "Ecological footprint – student calculator" by WWF. A discussion about the political rationality of these calculators follows, meaning how that rationality becomes the truth and what is possible to say and do inside that specific rationality. In that discussion we return to our question about the roles of external actors in education and what kind of influences they might have.

8.4 "My Climate Footprint" by E.ON

The calculator, "My climate footprint" (*mitt klimatfotspår*), is produced by E.ON, the largest private energy company in Sweden. E.ON produces and delivers energy to the Nordic market in the form of electricity, gas, heating, cooling, waste treatment and energy-related services to approximately one million customers. On E.ON's website, https://www.eon.se/om-e-on/om-foeretaget0/foeretagsfakta.html, a special page called "Energy knowledge in school" (*Energikunskap i skolan*) targets students and teachers in school and provides a range of educational resources. Referring to their engagement in schools, E.ON states the rationale for engaging with young people and teachers in school: "That E.ON is a natural part of young people's consciousness when we talk about energy and sustainable solutions, creates a value for us, both now and in the long term." Thus, the reasons for providing educational resources is not only about helping school, but also marketing aspects of their engagement (cf. Andrée and Hansson, manuscript).

The resource of "My climate footprint"[1] targets both young people in school and (potential) customers. The design of the calculator follows the visual appearance of the E.ON website and logo at large and uses the same design idiom including colour scheme, font and overall website design. It calculates one's climate footprint, which is stated in number of kilograms of CO_2, and compares one's individual footprint with that of an average Swede and an average world citizen. When entering answers

[1] The calculator was previously found under the heading "teaching materials" on E.ON's website. However, it has now been moved from the main pages to http://www.eon.se/upload/eon-se-2-0/flash/om_e.on/miljo/klimatfotspar/eon_footprint.swf

into the calculator, the person taking the test is presented a slideshow which includes the following questions: "It's people who are causing climate change. How are you contributing? Is your impact larger or smaller than the average Swede? And how do you compare with the world citizen? "My climate footprint" gives you the answers in 5 min." Also stated in the introduction are the factors which influence the result, for example, how one travels, how one's home is heated and how much energy one uses. One is also given the information that the average Swede has an environmental footprint corresponding to 6200 kg CO_2 per year.

8.4.1 Questions and Categories

The E.ON ecological footprint calculator calculates the impact of an individual's consumption in terms of the amount of carbon dioxide emissions resulting from activities in four sections of questions: flights, daily travels, heating and electricity use. The bookkeeping required asks for a high level of detail, although it is restricted to only those four areas; for example, there are no questions about the consumption of clothes, food, electronics, et cetera. Instead, this type of consumption is added automatically as a standard value at the beginning of the test and is set to 1000 kg per person per year and is thus not possible to influence the person taking the test.

The persons taking the test are expected to keep a detailed journal of all areas of their energy consumption. One has to account for yearly travel in kilometres with respect to the distances travelled by airplane, car, bus and train. This is clearly asked in the example, "Daily travels: How far do you travel by car per year?"

The following response options are given:

- I never travel by car
- 1–5000 km per year
- 5000–10,000 km
- 10,000–15,000 km
- 15,000–20,000 km
- 20,000–30,000 km
- More than 30,000 km per year

Among the response options, the option 10,000–15,000 km is marked with an icon symbolising the distance travelled yearly by the average Swede. Similarly constructed response options, along with the value for the average Swede is given for all questions. Thus, what can be perceived as "normal" is always available for us to estimate our own individual consumption in relation to that of others.

The questions are directly focused on energy use – the core of E.ON production – and include questions of heating and electricity. The first question asked is if you would describe yourself as "thrifty", "conscious", or "comfortable" (*sparsam, medveten eller bekväm)* in respect to your use of energy in general? In addition, more specific questions are asked such as: How large is your living space and how many persons share that space? What room temperature do you have at home? How old is your fridge? How much time do you spend in the shower each day? These questions, and how they are posed, constitute a governing technology where the aim is to focus individual attention on specific everyday habits, pointing to the habits as a matter of choice and thereby influencing individual choices. The questions point to what aspects of one's life one should focus on, and the alternatives listed give one an indication of what is a supposedly good or bad answer.

Most questions carry an air of objectivity and focus on questions and numbers indisputably relevant to our ecological footprints. However, in a few questions, E. ON appears as an actor providing an alternative for individual action. One example is with the question of how one's home is heated. The alternatives given include electricity, district heating, district heating from E.ON and don't know. If you chose electricity, you are asked if you know how your electricity is produced with the following alternatives:

- Yes, I'm an EON customer and therefore have hydroelectric power
- Hydroelectric power
- Wind power
- Nuclear power
- "*Bra miljöval*" [an eco-label translated as good environmental choice]
- No [do not know]

In the response options for the above questions, the marketing function of the calculator is undisguised. However, in the previous questions, the virtues of E.ON were more implicit. The questions posed, the response alternatives, and also the advice on how to decrease one's personal environmental debt all put light on specific aspects of the ecological footprint. For example, in the heating and electricity sections, you get additional, more in-depth information about the advantages of district heating. The headline reads "District heating – a climate hero". Unsurprising is that district heating is a service which E.ON provides. When we examine our ecological footprints (yearly emission of carbon dioxide) by varying our responses, we find that if we state that we use electricity from E.ON as our source of energy for heating, then what room temperature we have set, what the size of our living space is, how old our fridge is, et cetera, does not at all affect the test results. One can say that by simply choosing electricity from E.ON, we decrease our personal environmental debt immediately and substantially and without making any other changes to our lifestyle. However, a note is made at the end pointing to the possibility of also helping to decrease the total human climate footprint: "Even if your electricity comes from environmental friendly water power, you can make a climate effort by saving energy", and "If you decrease your use of electricity, you free environmental friendly water power".

8.4.2 Numbers and Illustrations

When answering the questions in the calculator, we have to choose among different alternatives. One alternative is always marked as the "average Swede". This gives direct information about whether one's choice makes a bigger or smaller environmental footprint than others. When we took the test, we noticed that this arrangement had an emotional impact on us. We started feeling guilty when answering some of the questions, whereas when responding to other questions, we began to feel proud and a bit like environmental heroes. The emotional experience is amplified by a metre with a movable pointer, which summarises your ecological footprint at the end of each section. Through the metre, it becomes obvious for each section of questions whether you are below average, average or above average in respect of CO_2. Thus, although the calculator provides a tool for making judgements seemingly detached from feeling, passion and non-reason (cf. Bennett 2001), its governing becomes fundamentally emotional (Ideland and Malmberg 2015).

The final climate footprint – in carbon dioxide weight– is compared both to the average Swede and to the world citizen. We find that our results were well below the average Swede: 5 tonnes compared to 6.7 tonnes for the average Swede. And, more surprisingly, if we use energy from E.ON, we are almost on par with the world citizen: 4.6 tonnes compared to 4.3 tonnes for the world citizen. The results leave us with the feeling of being fairly good and slightly better than others. The "My climate footprint" calculator thus operates as a governing technique of comparing oneself to others; the disciplined, rational and responsible citizen is one that does more good than the average citizen. In a final comment, E.ON makes a note about solidarity and informs us that we can do more if we want to:

> Want to make a climate effort? You can reduce your carbon-dioxide footprint fairly easily by changing your habits. At the same time, you may be required to make greater efforts such as changing heating methods. We hope we can give you a little help on the way.

The website then directs us to two more teaching resources provided by E.ON, leaving us with a sense that we are doing fairly good, but if we want to do even better, we can put our trust in E.ON.

8.5 WWF Ecological Footprint Calculator

The other calculator is the ecological footprint calculator produced by World Wide Fund for Nature (WWF). WWF is an international, nonprofit environmental organisation funded in the 1960s with the aim of conserving nature in the former European colonies of Africa and Asia. Their concern was that international areas of natural beauty risked being exploited in the new geographical landscapes. Even today, WWF emphasises the conservation of species and lands in mainly exotic places (seen from a European perspective). However, since the 1990s, education for sustainable development, has become focused areas for the organisation. Education is thus recognised as "vital if people are to acquire the knowledge, skills, attitudes and values that they will need if they are to change the way they live in relation to conservation, biodiversity and other issues of sustainability" (Fien, Scott, and Tilbury 1999, p. 27). This is especially the case in Sweden where WWF Sweden calls itself one of the most prominent national WWF organisations when it comes to educational issues (Ideland and Tröhler 2015). They have a deep engagement in developing teaching materials suitable for all school years from pre-school to upper secondary school.[2]

One example of a learning tool WWF offers for emphasising the connection between "your" (or "my") lifestyle and the degradation of the global environment is the calculator for ecological footprints (http://www.wwf.se/utbildning_kalkylator/?page=1&PHPSESSID=ecd716bd5b44d33499e329774c8f85c2). Through this, either the school class or individual students (or indeed, anyone) can calculate how many global hectares his/her/their lifestyle requires. A global hectare is a unit for measuring how much biologically productive land and sea space is needed for producing and taking care of all that is consumed. On the web, one can find many calculators of ecological footprints aimed at individuals, countries, organisations, companies, et cetera. What they all have in common is that they illustrate the result through the metaphor of the planet: How many planets are needed if "everyone lived like you" or, as WWF puts it on their Swedish website,

> Everything humans do has an impact on our environment in some way. The food we eat, the clothes we wear, everything that is produced has an impact on the world's forests, oceans, rivers, soil, air, plants and animals. The more we produce and consume, the more living things around us are affected. When we study our ecological footprint we understand more about the impact and can contribute better to society's transition towards more responsible production and consumption.
>
> (http://www.wwf.se/vrt-arbete/ekologiska-fotavtryck/1127697-ekologiska-fotavtryck Our translation)

8.5.1 Questions and Categories

The student ecological footprint calculator accounts for the impact of the individual's food, energy, transport, clothes/electronics consumption and waste. These are the areas in which the student is supposed to keep a book, and it should be done thoroughly. The following are examples of numbers that are to be specified in exact quantities and put into the equation:

- What is your weekly consumption of meat, cheese, eggs, grain products, fruit, etc.? The calculator also asks for weight or quantity, and then to which extent these are produced in Sweden, and respectively, organic.
- How big is your home in square metres? How is it heated? How many kilowatt hours are spent aside from this, and do you buy eco-labelled energy?

[2] For a wider and deeper analysis of WWF's educational turn, see Ideland and Tröhler, 2015.

- How many kilometres a year do you go by car, train, airplane, bus, etc.? (Comment: No questions are asked about what kind of fuel)
- What is your yearly consumption of underwear, trousers, t-shirts, dresses, coats, shoes, etc.? All amounts are to be specified in respective quantities.
- How many kilos of different kinds of waste, such as paper, glass, metal, electronics, organic, plastic, etc. are produced? Also state what percent are recycled, deposited or combusted in each category.

Going through the calculator requires the student to be quite knowledgeable of his or her family consumption. Most likely, she or he would need to do some kind of study at home and engage their parents, or guardians, in the work. We, three educated adults with an idea of ourselves as aware of the relation between sustainability and our own lifestyles, had difficulties in accomplishing the task. We had to refer back to electricity bills and had a hard time estimating the weight of both the waste we put in our waste bin and left at the recycling station. It is certain that we were not exact in our estimations of our impact on the environment, but the important aspect here is not our result, but what type of numbers were kept in the book of ecological footprints and what these numbers do to set the standards for the sustainable citizen. How are these made up inside a political rationality and then manifested in the political programme of WWF?

The numbers accounted into the calculator are all related to everyday family life – food and clothing, transport and heating, and not least, waste. From these categories, two questions appear: What does it mean that these particular areas are in focus, and what other types of consumption are left out of the calculation? First, the consumption measured (with the exception of clothing) is related to family life (food, energy, transport, waste). On the one hand, one can question how the student can have an impact, but on the other hand, the students are also seen as the connection between political will (sustainability) and a family lifestyle. This is nothing new; throughout history the connection between society, family and school has been strong, with school having been used as a tool for fostering citizens in a specific political rationality (Popkewitz 2008). Second, the precision required for the numbers to be calculated (e.g. how many kilos of plastic do you recycle/deposit/combust every year?) opens up for a specific kind of governing where every little detail counts. Every scrap of paper and plastic container seems to impact one's ecological footprint as well as every transport and every square metre in one's residence. In comparison with the E.ON calculator, WWF does not provide any response options to the questions asked which means that when we take the test, we are not given any clues concerning the possible range of consumption in the different areas. Everything counts. And we are held accountable for keeping track. In this detailed counting, we are left with a sense of not being very knowledgeable.

8.5.2 Numbers and Illustrations

All areas of questions require numbers. Some of them were quite easy to answer for an adult (size of living area) while others needed some investigation (use of electricity in kWh), but most of them demanded an estimation: How much plastic waste do you (as a person, not a family) produce et cetera? It took a long time and many guesses to accomplish the calculation. The most difficult to estimate was the waste, but in the end, it made little impact on the results of the ecological footprint (4% of the value). This is an illustrating example of the elevation of exact numbers in the calculator; the student is required to engage deeply and for long time in the category of waste – which gives a sense of its importance. But in the end, its significance is almost negligible. Concerning the food consumption, efforts were put into estimating the amount of organic and locally produced food consumed, but when changing from a high amount of organic food to none, the difference in the result was quite insignificant. It also did not matter if one doubles their use of electricity. Consumer choices which seemed highly important in the questions proved to have little significance for the result.[3] Nevertheless, advice given from WWF often focuses on these specific areas: Buy organic! Lower the temperature! Recycle!

On the other hand, almost invisible numbers show up at the end of the test which are not possible for the individual do anything about: societal ecological footprints including public emissions and other environmental impacts from infrastructure, public buildings, healthcare, national defence – "*your share* of public consumption" (italics in original). Besides that, it is stated that the calculator adds numbers for other kinds of utilities such as furniture and books.[4] The global hectares and CO_2 emissions coming from these are more than half the "individual" ecological footprint. However, these are quite invisible even if they have a great impact on the result. Their invisibility, in combination with the elevation of the individual choices, thus direct the governing to the individual consciousness and point out the necessity of being constantly aware.

Through the aesthetics of the calculator, it is obvious that it was created for children and young people. Also, one finds images recognisable from WWF, with the famous symbol of the panda and the text "WWF©. For a living planet®" (English in original) being most prominent. The brand provides a certain kind of legitimacy for the calculation of the ecological footprint, both when it comes to its purpose (the living planet®) and its interests (a famous environmental organisation©). The planet is also used in the calculator as an illustration and as a point of comparison. Here, one's ecological footprint is translated into a specific number of planets. In our case, it was 3.6 planets (illustrated with a drawing of 3.6 planets).

---

[3] One exception here was flights in the category of "transport", which had a significant impact on the global footprint.

[4] The latter are interesting from a moral perspective: What types of supplies are considered possible to negotiate and consume less or differently (food, clothes), and what types of supplies are not up for discussion (books)? Food and clothes are also subjects related to other teaching materials published by WWF.

Here, the planet is used as a metaphor for the limited resources that mankind has at its so-called disposal (no other species are counted), which has been a common discourse in the environmental movement ever since the 1960s when environmental awareness increased (Höhler 2015). The notion of the planet and the global footprint thus operates as a governing technique through both calling for the preservation of nature and solidarity with others. A total of 3.6 planets would be needed if *everyone* lived like you.

The third prominent image in the calculator is the footprint, drawn as a foot containing things from the categories measured (a house, a plane, fish, some waste, a wind power plant, etc). The image illustrates the choices that can be made to impact one's footprint.[5] However, the footprint only contains the choices that are measured in the calculator, not the add-on from public consumption. Finally, besides the illustrated number of planets, what is important for understanding the calculation is a table of the numbers and the impact of every category. When clicking on the categories, one is directed to WWF's page for notices and fundraising. The message is clear: Engage yourself! This engagement can be materialised through various gifts to the organisation. Through the table of numbers, the stated – and objective, in terms of supposedly exact calculations – problems with one's consumption are translated into a solution: engagement through financially supporting WWF.

8.6 Governing Technologies of Bookkeeping, Debt and Guilt

To summarise, the self-calculating human being (Miller 2004), constructed through the two different calculators, makes the individual's environmental footprint visible and governable. This is done through a kind of bookkeeping of the consumption of energy, transport, and in the case of WWF, also food, clothes and waste. Individual habits in everyday life are measured, and through the calculators, these habits are translated into precise values, given in kilograms CO_2. For instance, one result from WWF was 9831.4 kg CO_2 (which included all the different categories). A figure for national public debt was added to this, as well as a figure for the estimated individual consumption of other goods (furniture, books, etc). Together, this gave a result of 16476.4 kg CO_2 which was translated into humankind terms using the resources of 3.6 planets (if everyone lived like the test person). The final result becomes an assessment of an individual's environmental footprint. However, how these translations are calculated is not transparent but rather remain an opaque algorithm.

The calculator can be compared to a black box hiding the processes of accounting (Latour 1999). Which algorithms calculate the footprints (Hansen 2015)? How do they translate everyday habits into exact numbers of environmental degradation? What are the most important, and respectively, less important variables? A lack of

[5] This foot shows up in many places in WWF's teaching material and is filled with different symbols: for example, a flag to symbolise the size of a national footprint, children of different colour to symbolise the common effort for the world, etc.

transparency is hidden behind seemingly exact values for a person's impact on the global environment. Detailed questions of limited areas of life are at the forefront, and thus, apparently important. However, many questions matter much less than others. This is not stated outright, but was recognised while we made changes in the bookkeeping. For instance, transport habits have much more impact than buying organic food according to WWF calculator (despite the many and detailed questions about the latter). However, according to the E.ON calculator, the choice of energy source (rather than the amount of consumed energy) trumped all other choices. Interestingly, using EON's hydroelectric power or district heating made it possible to decrease one's individual footprint dramatically to a level close to or even below the average world citizen.

One difference between the calculators is the point of comparison. If we stay in the economics metaphor, WWF provides humanity with a "budget" – a single planet. This metaphor is also used in other WWF campaigns. In August 2015 WWF Sweden stated that the world had reached "overshoot day" – the day when humanity has exhausted the budget for its yearly resources. During the rest of the year, humanity will maintain and increase its ecological deficit in the "planet budget" (Ideland 2017). In this way, WWF provides an *absolute* point of comparison for the individual environmental footprint (and consequently, the individual's lifestyle). You can either fulfil the standards for an environmentally friendly life or you fail and become indebted to the planet. On the other hand, E.ON provides *relative* points of comparison for the individual: the average Swede and the world citizen. Through E.ON's calculator, the standards for the sustainable citizen are measured against an average rather than against planetary resources. This means that as long as you behave better than others, the calculator does not individually point you out as a problem. This is especially manifested in the scale visualising individual carbon dioxide emissions which shows if one is below, over or average.

Through the technologies of accounting, both calculators produce indebted persons who are influenced to become aware of their part of the world's environmental problems in order to live up to the standards of a responsible person (Millar 2004). The debt arises from an individual's habits in everyday life, from buying shoes, eating, heating one's house and taking care of waste. In the case of the WWF calculator, figures for the national public debt (although not emphasised) are also added to the individual result. In contrast, the calculation from E.ON did not include the national public debt; the individual and his or her choices thus become even more important in this particular test. As Lazzarato (2012) claims, the indebted human being is a prerequisite for the neoliberal rationality in which every choice is elevated as important. Through this debt, individuals need to comply to a public will in order to pass as responsible persons – their lives become administrable, and all the details accounted become possibilities for change.

Through the numbers, their accuracy and the bookkeeping into debt, the calculators produce a rational ethical approach (Rose 1991). Becoming a more ethical, responsible person is done through changing one's lifestyle. The conscious choice of almost everything is viewed as a way to escape the debt and do good for the world. To some extent, this is communicated through explicit advice, but also indirectly through the way in which the questions are posed in the tests. However, this rational approach to ethics is also combined with a more emotional approach – the debt is translated into guilt.[6] Overall, guilt operates as an important node in sustainability and environmental movements, and thus, also in education. Through the individualisation of environmental problems, personal guilt is knitted together with global threats, and detailed individual activities are described as possibilities for rescuing the flock and the planet (Ideland and Malmberg 2015; Lazzarato 2010). Thus, guilt becomes an important emotion for the governing of human beings and possibly a prerequisite for targeting the consciousness. It is a pastoral type of governing which has been frequently used throughout history, especially within religious institutions (Foucault 1983; Ideland and Malmberg 2015; Ideland forthcoming). This pastoral power is most obvious in the calculator by WWF, where the budget for humanity is clearly stated (one planet). The E.ON calculator offers an easy path to salvation – through changing one's energy company. Thus, the debt – translated into guilt – becomes an ethical technology.

8.7 Neoliberal Rationality Through External Actors

When the calculators are put in the hands of students as part of science education or education for sustainable development, the calculators contribute to the making of a neoliberal political rationality where so-called rational solutions to our environmental problems are mainly individual ones. The individual student is invited to scrutinise her or his choices in everyday life. However, in a different political rationality, the solutions to our environmental problems would primarily be stated as political decisions, for example, by way of stricter legislation. From such a political rationality, the individual should be encouraged to act as a citizen instead and try to achieve change through political decisions (e.g. which energy sources are used in the society, regulations for how food should be produced, and the cost of various modes of public transportation).

The discourse of helping school to solve "the science education problem" opens up for economic and ideological interests to enter the classroom without being critically scrutinised. In the distributed statework, external actors such as industrial actors and NGOs become naturalised. In the two cases analysed in this chapter, we have seen how the distributed statework opens up for a neoliberal rationality to enter the classroom. It is of the utmost importance that teachers and educational policymakers are made aware of the governing elements within the different types of teaching materials provided by external actors.

Our intention with this chapter is to contribute to the close examination of how political rationalities are invited into classrooms through various types of external actors. We have exemplified this by analysing two "ecological footprint calculators"

[6] In Swedish, the word "skuld" means both *debt* and *guilt*. It is the same word which makes the transformation almost invisible.

designed by E.ON (a private energy company) and WWF (a non-profit environmental organisation) as resources for teaching about energy use and resources. We have argued that both calculators produce indebted and guilty individuals – but to different extents. While E.ON's calculator measures the individual result against an average Swede and an average world citizen, WWF measures the individual result against one's part of the globe. This makes the illustration of the debt vastly different in the two calculators. However, the types of solutions offered are similar in that they are consequences of a specific political rationality which correspond with neoliberal individual solutions.

Indeed, analysing and understanding different interests and political rationalities (e.g. in terms of suggested solutions to environmental problems) could be a learning goal in its own right for science education. These types of materials from an array of external actors could be used as resources to enrich the teaching of science; for example, in regard to the footprint calculators, a major challenge is for teachers to highlight and uncover the rationalities of the calculators which are hidden by the use of numbers and non-explicit advice. One way would be to begin asking simple questions about what is included in the resources and what is not. What is accounted for and how? What are the reference points? And what interests are potentially being opened up in the science classroom when opening the classroom doors for a particular actor? We acknowledge that it may not be easy to balance the mission of providing a so-called objective education and the mission of enriching education by involving the surrounding community. Further, this becomes even harder when the political rationalities present in the external resources are well in line with dominant political rationalities; in this case, the rationalities of individual accountability and individual choice (cf. Ball 2009). The analysis presented in this chapter may contribute to raising teachers' and science educators' awareness of the tensions that, by necessity, will follow the opening up of the teaching of science to contributions from external actors with specific (more or less clearly stated) interests.

References

Andrée, M., & Hansson, L. (2014). Recruitment campaigns as a tool for social and cultural reproduction of scientific communities: A case study on how scientists invite young people to science. *International Journal of Science Education, 36*(12), 1985–2008.

Andrée, M., & Hansson, L. (2015). Recruiting the next generation scientists and industrial engineers: How industrial actors engage in and motivate engagement in STEM initiative. (symposium International Organization of Science and Technology Education, IOSTE, Kuching, Malaysia, September 2014). *Procedia-Social and Behavioral Sciences, 167*, 75–78.

Andrée, M., & Hansson, L. (manuscript). *Industrial involvement in STEM-education: Actors, resources and repertoires*.

Ball, S. J. (2009). Privatising education, privatising education policy, privatising educational research: Network governance and the "competition state". *Journal of Education Policy, 24*(1), 83–99.

Bennett, J. (2001). *The enchantment of modern life: Attachments, crossings, and ethics*. Princeton/Oxford: Princeton University Press.

Eriksson, M. (2016). *Berättelser om Breivik. Affektiva läsningar av våld och terrorism. [stories about Breivik. Affective readings of violence and terrorism]*. Göteborg: Makadam. Diss.

Foucault, M. (1983). The subject and power. In H. Dreyfus & P. Rabinow (Eds.), *Michel Foucault: Beyond structuralism and hermeneutics* (pp. 208–226). Chicago: University of Chicago Press.

Foucault, M., Senellart, M., & Davidson, A. I. (2007). *Security, territory, population*. Basingstoke: Palgrave Macmillan.

Fien, J., Scott, W., & Tilbury, D. (1999). *Education and conservation: An evaluation of the contributions of educational programmes to conversation within the WWF Network*. Washington, DC: World Wildlife Fund.

Goldman, D., Assaraf, O. B. Z., & Shaharabani, D. (2013). Influence of a non-formal environmental education Programme on junior high-school students' environmental literacy. *International Journal of Science Education, 35*(3), 515–545.

Hacking, I. (2006). Making Up People. *London Review of Books, 28*(16), 23–26.

Hansen, H. K. (2015). Numerical operations, transparency illusions and the datafication of governance. *European Journal of Social Theory, 18*(2), 203–220.

Höhler, S. (2015). *Spaceship earth in the environmental age, 1960–1990*. London: Pickering & Chatto.

Ideland, M. (2017). The end of the world and a promise of happiness: Environmental education within the cultural politics of emotions. In T. Popkewitz, J. Diaz, & C. Kirchgasler (Eds.), *A political sociology of educational knowledge. Studies of exclusions and difference*. New York: Routledge.

Ideland, M., & Malmberg, C. (2015). Governing "eco-certified children" through pastoral power: Critical perspectives on education for sustainable development. *Environmental Education Research, 21*(2), 173–182.

Ideland, M., & Tröhler, D. (2015). Calling for sustainability: WWF's global agenda and teaching Swedish exceptionalism. In D. Tröhler & T. Lenz (Eds.), *Trajectories in the development of modern school system. Between the national and the global*. New York: Routledge.

Ideland, M., Axelsson, T., Jobér, A. & Serder, M. (2016). *Helping hands? Exploring school's external actor-networks*. Paper presented at ECER conference, Dublin 2016.

inGenious (2014). *What is inGenious?* (Leaflet). Retrieved 12 Feb 2014, from http://www.ingeniousscience.eu/c/document_library/get_file?uuid=bb4ed239-ec5b-47b2-b8ae-4ba96500ab83&groupId=10136.

Latour, B. (1999). *Pandora's hope: Essays on the reality of science studies*. Cambridge, MA: Harvard University Press.

Lazzarato, M. (2010). Pastoral power. Beyond public and private. *Open, 19*, 18–32.

Lazzarato, M. (2012). *The making of the indebted man: An essay on the neoliberal condition*. Los Angeles: Semiotext(e).

Miller, P. (2004). Governing by numbers: Why calculative practices matter. In A. Amin & N. Thrift (Eds.), *The Blackwell cultural economy reader*. Oxford: Blackwell Publishing.

National Agency for Education [Skolverket]. (2011). *Curriculum for the compulsory school, preschool class and the recreation centre 2011*. Stockholm: Swedish National Agency for Education.

Popkewitz, T. (2004). The alchemy of the mathematics curriculum: Inscriptions and the fabrication of the child. *American Educational Research Journal, 41*(1), 3–34.

Popkewitz, T. (2008). *Cosmopolitanism and the age of school reform. Science, education and making society by making the child*. New York: Routledge.

Rose, N. (1991). Governing by numbers: Figuring out democracy. *Accounting, Organizations and Society, 16*(7), 673–692.

Rose, N. (1996). Governing "advanced" liberal democracies. In A. Barry, T. Osborne, & N. Rose (Eds.), *Foucault and political reason: Liberalism, neo-liberalism and rationalities of government*. London: UCL Press.

Rose, N. (1999). *Powers of freedom: Reframing political thought*. Cambridge, MA: Cambridge University Press.

Rose, N., & Miller, P. (2010). Political power beyond the state: Problematics of government. *The British Journal of Sociology, 61*, 271–303.

Stables, K., & Kimbell, R. (2001). Technology education in South Africa: Evaluating an innovative pilot project. *Research in Science Education, 31*(1), 71–90.

Teknikdelegationen (2010). *Engagemang för naturvetenskap och teknik – fördjupad kunskap om svenska initiativ som syftar till att öka barns och ungdomars intresse för ämnena.* Rapport 2010:3. Retrieved 20 Mar 2014 from http://www.skolverket.se/polopoly_fs/1.129012!/Menu/article/attachment/Rapport%25202010_3.pdf

Chapter 9
What's in It for Me?: How Does a Professional Development Programme Meet Science Teachers' Career Expectations?

Peer S. Daugbjerg and Martin Krabbe Sillasen

9.1 Background

A professional development programme (PDP) is a systemic entity where different stakeholders work together with the aim to create optimal opportunities for teachers' professional development. For many years, researchers in teachers' professional development have investigated the relationship between programmes and the persons involved in science teacher professional development (Hewson 2007). An important point from Peter Hewson's review (2007) is that the role of programme structures is facilitative not causative. Teachers themselves are responsible for changing their practice and in the process empowering themselves. Considerable efforts have been made into not only addressing how PDPs are successfully implemented but also the relationship between teachers and PDPs. There is a growing awareness that a "one-size-fits-all" PDP does not meet contemporary requirements for professional development. Programmes must be designed to adapt more effectively to an individual teacher's needs (Luft and Hewson 2014).

Even though many PDPs are designed to accommodate the individual professional learning needs of teachers, we have elsewhere questioned whether these types of PDPs actually improve teachers' learning effectively (Daugbjerg 2015; Daugbjerg and Sillasen 2016) because teachers' individual career expectations affect how the teachers relate to a given PDP. In this chapter, we will continue this line of inquiry by troubling whether PDPs that are designed to accommodate and

P.S. Daugbjerg (✉)
VIA University College, Teacher Education, Lemvig, Denmark
e-mail: pd@via.dk

M.K. Sillasen
VIA University College, Teacher Education, Silkeborg, Denmark
e-mail: msil@via.dk

K. Otrel-Cass et al. (eds.), *Cultural, Social, and Political Perspectives in Science Education*, Cultural Studies of Science Education 15,
DOI 10.1007/978-3-319-61191-4_9

improve individual teachers' professional capacity really are more effective. Often PDPs prescribe what teachers should be like and how they can become these desired professionals possessing the pedagogical content knowledge necessary for teaching a subject (Shulman 1986; Gess-Newsome and Lederman 1999). Teachers' subjective experience of a PDP is, however, an assemblage of their own life history, professional career, present educational politics and discourses (Webb 2009). Accountability is one way of framing the desired teachers in educational politics (Webb 2009), and PDPs can be considered as a tool to implement educational reform policy (van Driel et al. 2012). Taylor Webb (2009) analysed the idealised micropolitics imposed on teachers' meaning making using accountability discourse. His analysis shed light on how any given PDP also contains a discourse of idealised meaning making for the participating teachers. In the research we present here, we investigate how teachers experience PDPs and whether they believe PDPs provide opportunities for developing their teaching practice and future career.

We address the perceived opportunities through the notion of *foreground*. The notion of foreground is developed empirically within mathematics teaching to describe pupils' dispositions for engaging in learning and teaching. Foreground sums up all the dynamic relations between the pupils' lived experience and their expectations for the future (Skovsmose 1994). Alrø et al. (2009) further explicated foreground as expectations of future use and benefit of current learning. Such expectations address how particular teacher groups make meaning of the ways a given PDP offers to change their own teaching practice.

A mixture of one's own experience *and* the possibilities for capacity building offered by the PDP comprise the conditions for the teachers' participation (Daugbjerg 2015; Daugbjerg and Sillasen 2016). However, addressing science teachers' participation in PDP in this stereoscopic manner is not possible using a single approach, because we need to juxtapose the systemic PDP ideals with the teachers' individual career expectations. We have therefore combined two different approaches: one from the systemic point of view of the PDP and one from the individual point of view of the participating teachers. In order to honour the different characteristics of these two approaches, this chapter is organised as a theatre play in three acts, each with a different scene. The first act presents the PDP point of view with applied methods and generated findings; in this act the research and analysis were primarily performed by Martin Sillasen. Sillasen and other researchers within the framework of the QUEST PDP (Nielsen and Sillasen 2014) designed this research. The second act presents the teachers' point of view with applied methods and generated findings; in this act Peer Daugbjerg primarily performed the research and analysis. Sillasen and Daugbjerg collaborated on designing the research from teachers' point of view, but Daugbjerg performed the research. In this chapter, we bring together the two different perspectives on PDP ideals of QUEST, and the micropolitics and foregrounds of the teachers' participation in the QUEST PDP. We are thus situating and communicating our interpretation very deliberately from two distinct perspectives. This joint stereoscopic approach is inspired by the notion of "research as bricolage" put forward by Kathleen Berry (2015) and Shirley Steinberg (2015). Our bricolage – a joint stereoscopic approach – helped us to centre the participating teachers in our research in order to provide a more saturated appearance of their experience with QUEST (Berry 2015).

In the third and final act, we bring together and discuss these two points of view on science teachers' participation in PDP. In this act, the two research approaches are merged, and both authors have performed the entangling analysis presented here. This final analysis leads us to discuss what can be learned about a PDP, depending on how teachers' learning outcomes are assessed. In what follows, we address how teachers' individually perceived foreground relates to the "ideal anticipated foreground" that a Danish PDP – QUEST – is based on in order to optimise teachers' outcome.

9.2 First Act: Characterising How to Model Participating Teachers' Anticipated Foreground in a PDP

The Danish PDP called "qualifying in-service education of science teachers" (QUEST) is intended to balance top-down and bottom-up processes in order to integrate collaborative and individual perspectives of the participating teachers in the overall programme (Nielsen and Sillasen 2014).

QUEST activities are based on five research-informed didactical principles, which encompass the systemic interpretation of *an anticipated foreground* that participating teachers experience. The didactical principles (QUEST 2012) of QUEST are:

1. Improving student learning is the central aim for all PDP activities.
2. Teachers are active in PDP activities.
3. Teachers experience cohesion between their learning needs and PDP activities.
4. PDP activities stretch over time to allow for experimenting with new teaching strategies in their own practice.
5. PDP activities take place in a collaborative and supportive environment.

The aims and activities were negotiated with the participating teachers before each course module to ensure alignment between teachers' foreground and the anticipated foregrounds prescribed by QUEST.

In this act, we as authors primarily aim to unfold design *characteristics* of the didactical principles that presented themselves through the realisation of course modules and present data on how these design characteristics were perceived by the participating teachers. The design characteristics that we investigate are:

- The QUEST rhythm (will be explained shortly)
- Alignment of teachers' expectations and QUEST aims
- Assuming different roles in a learning community
- Opposing the hampering factor of teachers' autonomy

The design characteristics in some sense prescribe the systemic foreground anticipated by QUEST to optimise the individual teacher-learning outcome. We have inquired into the overall picture by using:

- Questionnaires distributed to participating teachers. The questionnaires were distributed prior to and after (spring 2012 and autumn 2014) the realisation of the course modules. They contained five-point Likert scale questions and open-ended categories focussed on teachers' experience of course modules. Sillasen distributed these questionnaires.
- Individual and group interviews with participating teachers and municipal science education consultants. These interviews provided information about what hampered or supported successful professional development activities within the municipality. Sillasen performed these interviews.

These data provided information on participants' views on the systemic perspective of the PDP QUEST.

Our informants were from five municipalities in the central region of mainland Jutland in Denmark. They all teach primary science or a science subject in lower secondary school. One hundred and thirty-eight teachers participated in this part of the research; fifty-one answered a questionnaire.

In the following sections, the analysis of the four design characteristics presented earlier is elaborated and discussed. We however start by presenting more details on the QUEST programme and its research base in order to provide a saturated description of the setting where our research was conducted.

9.2.1 Research Context and the QUEST Rhythm

In QUEST, learning networks were organised across school boundaries within each municipality. The concept of professional learning network was used to characterise the collaborative activities that provided learning opportunities for the participating teachers (Jackson and Temperley 2007; Sillasen and Valero 2013). Professional learning networks share many commonalities with school-based professional learning communities (PLCs) such as a focus on student learning, developing teacher collaboration and individual teacher learning. But their additional purposes include enlarging individual schools' repertoire of choices and moving teaching ideas and examples of good practice around between schools in order to help transform the whole school system within a municipality, not just the individual school, thus improving education for many students. This lateral capacity building (Fullan 2007) is a collective responsibility and moral purpose is writ large.

Even though a PDP like QUEST is carefully designed, using research-based knowledge poses a risk that the teacher culture might not be changed if only a top-down perspective is included:

> We find that system conditions that support the work of PLCs – such as a comprehensive education plan, integrated learning resources, local knowledge resources, robust data and accountability system, extended time for teacher collaboration, and leaders committed to PLCs – are not sufficient to engender change in professional culture and teachers' work lives. The literature point to goals for system change, but offers little guidance on the change process or warning of pitfalls and challenges entailed in changing professional culture from the top. (Talbert 2010, p. 556)

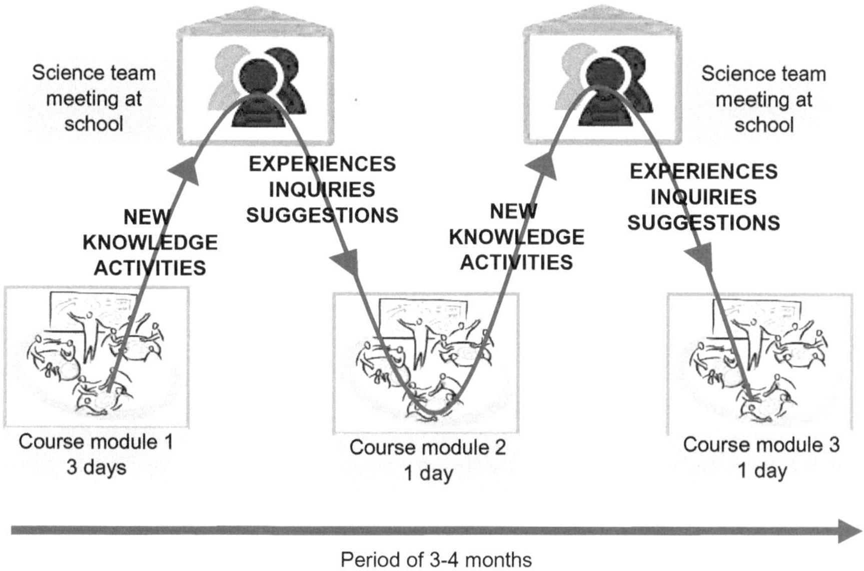

Fig. 9.1 The QUEST rhythm of course modules

What is needed to make PDPs, such as QUEST, more effective is to provide opportunities for teachers to experiment with teaching strategies followed by collective reflections. A significant feature called "the QUEST rhythm" was developed to accommodate the challenge of changing the professional culture in the participating schools. The "QUEST rhythm" describes a structure of the course modules, where the participating teachers alternate between workshop participation and working at their own schools: teaching, sharing and collaborating with colleagues (Fig. 9.1).

Teachers from different schools participated in the municipal workshops over a 2-year period (2012–2014), where they learned about new teaching strategies and materials. In the workshops, they also planned implementation of this new knowledge in their teaching practice. The teachers' learning process was supported by action learning to qualify their reflection on action and knowledge sharing. Following the implementation in their own classes, the teachers met again to share experiences and participated in collective reflections on their individual teaching experiences.

9.2.2 Findings from the Questionnaire and the Interviews

The participating teachers generally found that the QUEST rhythm supported their professional development (see Fig. 9.2).

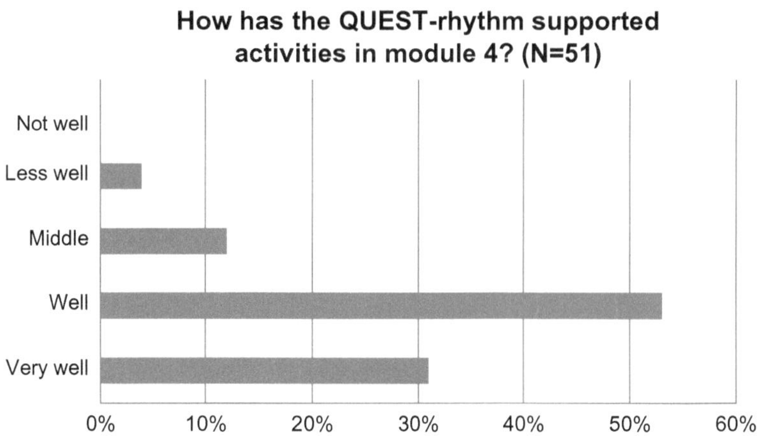

Fig. 9.2 Participating teachers' assessment of the benefits of the QUEST rhythm, alternating between course activities and working at their school

In interviews, teachers stated that through the interaction between receiving new teaching ideas in the workshops and being able to try them out in their own practice *and* interacting with peers in the action learning processes improved their professional learning. A teacher reflected:

> I think that the strength in QUEST is that we continuously have to experiment with the new teaching ideas in our own teaching. And then come back to subsequent workshops and present results…I think this process helps us to keep focus on our own learning process…it is a dynamic way of keeping focus over time. As participating teachers, we have two tasks in QUEST: Learning new teaching ideas and reflecting with peers on how it works in practice.

The participating teachers valued the knowledge sharing. Another teacher remarked:

> I think that we have never doubted the value of knowledge sharing. But now we have a structure (The QUEST-rhythm) that helps us…or in reality forces us to do what we always have thought would be nice to do…because now the school leaders have allocated time for knowledge sharing…and also because…well, in order to share knowledge about teaching qualitatively we need to trust each other.

9.2.3 Alignment of Expectations and QUEST Aims

Even though most PDPs aim at improving the professional capacity of teachers, researchers have argued that teachers might develop some kind of resistance towards top-down initiated activities:

> With all good intentions and research-based knowledge, district and school administrators sometimes create policies and routines that interfere with progress, and they wonder why teachers respond in unanticipated ways. Instead of jumping into collaboration with their

> colleagues, teachers sometimes organize to oppose new designs for their work or enact them in a routine fashion. (Talbert 2010, p. 556)

In QUEST, we experienced that participating teachers' expectations were not always aligned with the intention of action learning in the QUEST rhythm – as described above. Some teachers valued the collective reflections on experimenting with teaching activities. However, they only saw this process as secondary to receiving new ideas for their teaching. As one teacher argued: "Many were sceptical about the reflections on teaching experiences...But as we got started, a good debate on IBSE progressed...[But] as long as you get something that you feel you can use in your own teaching. I mean, I need an outcome from this collective planning and reflection that is concrete in relation to my own teaching. Otherwise I might feel that I could use the time more effectively". Another teacher argued: "I think that the best workshop activities were the concrete teaching activities that we could experiment with directly in our own teaching practice". However, there were also teachers who valued the workshops as they supported changes in their schools: "Exemplary workshop design with synergy between theory and practice. Good to share and try out new knowledge at own school, and then coming back getting new inputs and sharing experiences".

9.2.4 Assuming Different Roles in a Learning Community

The QUEST rhythm is designed so that teachers act as change agents with the purpose of changing the teaching culture within their schools. They do this by inviting colleagues to collaborate on planning teaching activities and knowledge sharing with their local professional learning community. The teachers' role as a change agent is vital for local implementation. However, some teachers consider the role challenging: "I have to say that I am challenged in conveying new teaching ideas to my colleagues...uhm, it is the most challenging part of participating in QUEST. And that surprises me...I cannot say why...when we are sitting in a group we are equal... but I find it difficult to take on the "expert-role". The role as change agent is challenging, because they are used to a teacher culture where everybody is equal. When they take on the "expert role", they step out of the "peer role". The "expert role" might come with uncertainty if the teachers experience a lack of response from other teachers in the PLC.

9.2.5 The Relation Between Teachers' Autonomy and the Intended Ideal of the QUEST PDP

In a teacher culture based on teachers' autonomy in planning and performing teaching, teachers doubted or even worked against the intentions of the QUEST PDP. This was the case in some of the participating schools. As Joan Talbert (2010) argued:

> The challenges of developing teacher collaboration are many. One stems from a tradition of autonomy in teaching that works against the formation of PLCs. When instruction is considered private practice, teachers resist the idea of collaborating with colleagues on instruction. They resist even more the opening of classrooms to peer observation and subsequent feedback (p. 557).

There were indications that several iterations of the QUEST rhythm over 2 years altered participating teachers' attitude towards collaboration about planning and evaluating teaching activities. A teacher argued: "I think that [collaboration] evolves over time…We do not come back from workshops and say that this and this teaching idea works in this and this way and then our colleagues just use it without some discussion. This autumn, my closest colleague and I have collaborated in new ways. We now meet and plan activities that each of us subsequently use in our own class. Before QUEST, we would never meet to plan teaching activities collectively. That has changed. Our collective planning and subsequent evaluation of teaching activities has improved our confidence in experimenting with new teaching strategies".

Despite the initial doubt about the possible outcome of participating in QUEST, slowly the teachers adapted the intended collaboration to their own local context.

9.2.6 The Significance of Collaboration and Networking

These findings indicated that collaboration and networking can be key elements for improving teacher learning, but the supportive effect depends on the degree to which the schools' organisational structures and leadership support and enable teacher commitment.

Talbert's (2010) concern about teachers opposing the changes in their practice intended in a PDP was somehow overcome in QUEST. As seen in the teachers' reaction to QUEST, there was more hesitance to immediately take over the intended ideal of collaboration, but when the existing culture was given time to accommodate to the collaboration ideal, then teacher collaboration developed.

In this first act, some of the design characteristics of QUEST were elaborated and discussed with the aim to characterise how QUEST has tried to accommodate the participating teachers' learning needs. The teachers experienced the collaborations as challenging but also rewarding in terms of creating a collective approach to planning and assessing teaching. We now shift scenes and assume an individual teacher's perspective, where a different perception of the learning opportunities offered by QUEST was revealed.

9.3 Second Act: Significance of Teachers' Career Expectations for Their Participation in PDP

Teachers' personal intentions for participating in PDP are formed by their lived experience and foregrounds (Daugbjerg 2015). The amalgam of personal experiences and foreground shapes the personal interpretative framework that each teacher uses to understand his or her actual challenges and problems as well as to engage in development initiatives (Kelchtermans 2009). Insight into the teachers' personal understanding of present challenges and problems in the classroom can inform professional development initiatives on how to succeed (Rolls and Plauborg 2009). QUEST attempted to meet this personal aspect of science teachers' PDP participation by including network establishment at local schools within a municipality (QUEST 2015). Formation of school-based professional learning networks has proved beneficial for creating a base for development of science teaching (Sillasen and Valero 2013).

Tensions in the relation between the personal and systemic intentions raise serious questions regarding whether it is at all possible to create successful PDPs as discussed in Sect. 9.2.3.

As indicated in the introduction, we are aware that the participating teachers also present a relevant point of view in our attempt to understand the phenomenon of science teachers' participation in PDP. The systemic perspective analysed in the previous section – act 1 – presented how collegial collaboration opens up joint work within planning, performing and evaluating science teaching. However, taking the individual teachers' perspective might provide a different story about teachers' learning outcomes from participating in QUEST activities. Acknowledging this left us as authors with the task to acquire insights into the participating teachers' personal point of view. The above-mentioned focus group interviews had given us some insight into the personal perspective, but we had to inquire deeper into this individual participant foreground perspective. We had to know more about individual teachers' background on why they became a teacher and specifically chose science teaching as their profession. We needed their individual reflections on their participation in QUEST in order to saturate our understanding of their individual point of view because we also sought to understand their teaching practice in detail.

This left us with the task of conducting in-depth interviews and observations with individual teachers. In order to capture the diversity of the participating teachers, we worked together with one teacher from each of the participating municipalities to create interviews and generate observations. These five teachers had overcome the first hard induction time of a new teacher, meaning that they were not struggling with issues typical for newcomers in the teaching professions. They participated in the QUEST courses and had experiences with the QUEST rhythm, and their schools had more or less well-functioning science teacher teams. One of these five left science teaching during the research period, so his data were not analysed. In this section, we present in-depth interview data from four teachers.

The interviews were conducted over two sessions; first an introductory interview was conducted with emphasis on the professional life history of the teacher; this interview was followed by whole-day observations of their teaching including video observation of one science lesson. The introductory interview and the observations were preliminarily analysed using NVivo software by Daugbjerg; this preliminary analysis generated a unique set of questions for each teacher for a final concluding interview. The complete data set was analysed using NVivo for the teachers' foreground (Daugbjerg 2015) and the teachers' participation trajectories (Daugbjerg and Sillasen 2016). In the individual interviews, the science teachers reported what they as individuals found interesting and relevant in QUEST. We now present four teachers' personal reflections on how elements of the QUEST courses related to their own everyday science teaching. All presented names in the following are aliases.

9.3.1 *Poul, 30 Years Old, 4 Years of Teaching Experience: Participating Science Teacher from Dandelion Municipality*

Poul had always wanted to become a science teacher since his youth, and he is dedicated to keep on teaching science. He is focussing on his present classroom teaching and on improving his science teaching.

Danish interview transcript	English translation
"Jeg vil fortsætte i det der med idedatabaser, rent kommunalt indenfor Mælkebøtte kommune, hvor hvis man havde nogle ideer og lagt dem ind, … hvor man så får noget feedback, 'den var smart, den vil jeg lige prøve af herude på skolen' og det synes jeg, er en rigtig god ide. På den måde kan det bruges"	"I will continue with these idea databases, strictly within the municipality of Dandelion, where you, if you have some ideas post them … where you get feedback. 'that was a good idea that one I will try it out at this school' and that I think is a very good idea, in that way it is useful"

Danish interview transcript	English translation
"Jeg blev mere gjort opmærksom på, at det man ser, og det man gør, det er altså, det man husker. …. Så er det måske mere vigtigt at vide, at indikatorpapir i cola bliver rødt, fordi det er en syre, end hvis de ikke kan huske alt muligt om H^+ ioner og alt muligt, som de ikke kan huske. Så er det første måske det vigtigste, fordi de kan huske det"	"I have been made more aware, that what you see, and what you do, that is really, what you remember. …. So it is perhaps more important to know that litmus paper in cola turns red, because it is an acid, than when they cannot remember every detail about H^+ and the like, that they cannot remember anyway. Then perhaps the former is the most important because that is what they remember"

Poul is clearly pleased with the design of QUEST and sees possibilities for developing his own science teaching and for contributing to developing his colleagues' science teaching within the municipality. Poul also says that he sees himself as a science teacher in the future. Poul is focussing on his present classroom teaching and on improving his teaching; his foreground is to continue teaching science.

9.3.2 Birger, 47 Years Old, 13 Years as a Teacher: Participating Science Teacher from Daisy Municipality

Birger changed to the teaching profession after 13 years as a hardware dealer. His interest for nature emerged in very early childhood; he has been studying nature and science literature all his life. He has ever since he left lower secondary school known that someday he wanted to become a teacher. During the interviews, he repeatedly talks about becoming a teacher educator. He senses that his approach to education can be better fulfilled as a teacher educator.

Danish interview transcript	English translation
"Jeg prøver også at få dem [eleverne] til at undersøge. QUEST ligger meget op ad det, som jeg også selv godt kan lide. ... Jeg bliver nødt at trække på, at de elever de har telefoner, og jeg bliver nødt til at trække på, at de synes det er sjovt. Fordi ellers kommer vi ingen vegne. Hvis jeg skal stole på, at de computere jeg booker [her på skolen], at de virker... De gør de ikke. Så mange har vi slet ikke. Når vi laver sådan nogle ting, så siger jeg:'vil I ikke være søde og rare at tage jeres egne computere med?'. Og det gør de, fordi de ved de virker"	"I try also to make them [the students] inquire. QUEST is close to what I also like. ... I have to rely on that the students have [smart] phones, and I have to rely on that they find it fun. Because otherwise we get nowhere. If I should rely on whether the computers I book [here at the school] work... They do not. We do not even have many here. When we do such teaching, I say: 'Please be kind and bring your own PC?' Moreover, they do, because they know they work"

Danish interview transcript	English translation
Peer:" Hvad forventer du er det største udbytte ved at deltage i et projekt som QUEST?" Birger: "At få nogle kolleger der er med på det. ... Men det er det der med, at vi er ikke så gode til at arbejde sammen. Og det kunne jo være dejligt. Det synes vi også, når vi har den her fag-fredag, så har vi faktisk ret stor succes med den. Og når vi først får sat os ned, så tager det faktisk ikke så lang tid at prøve at pejle os ind på, hvordan et undervisningsforløb skal ligge ift. det emne, vi nu gerne vil ind omkring"	Peer: "What do you expect to be the greatest outcome of participating in QUEST?" Birger: "To have some colleagues who take part in it. ... However, the thing is we are not that good at collaborating. But, that would be nice. We also feel when we have these subject-fridays, and then we do have a great success. In addition, when we do get together, then it actually does not take that long to tune into, how to teach a subject we want to focus on"

Birger finds QUEST relevant because its didactical position is similar to his own position. Generally, he acts autonomously in relation to supportive structures. He does not use the computers the school can provide, and he finds collaboration very close to everyday teaching rewarding. Birger expresses intentions of becoming a teacher educator. In the interviews, Birger also accentuates his devotion to evidence and an explicit knowledge base for science teaching. Birger's career foreground is outside lower secondary school.

9.3.3 Karl, 33 Years Old, 7 Years as a Teacher: Participating Science Teacher from Marigold Municipality

Karl has been engaged in a Danish Boy and Girl Scout movement [FDF] for more than 20 years, where he has been working with developing children's personal character and practical skills. Karl has been very focussed on becoming a science teacher and has taken a – for Denmark – special 3-year teacher bachelor training programme focussing exclusively on physics, chemistry and mathematics. In the interviews, he repeatedly talks about wanting to be a science teaching guide for his colleagues. Karl emphasises how collaboration with the school can support other colleagues to improve their science teaching. This emphasis relates back to his scout engagement on personal development and coincides with his aspired foreground as a science teacher guide.

Danish interview transcript	English translation
"… at bede folk om tage hjem og køre progressionstræer igennem, som er taget fra det her amerikanske ATLAS-projekt, som er super, som jeg er rigtig glad for at blive introduceret til. Det var et rigtigt spændende oplæg, de to undervisere lavede om det, jeg kunne vitterligt se noget idet. Og så bede os om at gå hjem og køre det igennem på skolen med nogle kollegaer på 2 timer for at få en ide om det, det rykker ikke og en meter"	"… to ask people to go home and make progression-trees, which comes from the North-American ATLAS project, which is super, and I'm very glad to be introduced to it. It was a very interesting presentation given by the two educators on that; I could really see something in it. Then they asked us to go home and work through it at the school with some colleagues in 2 hours to get an idea of whether it changes anything"

Danish interview transcript	English translation
"… men der sidder jo alligevel nogen [kollegaer] som (.) som sidder der af pligt ikke. Det er forståeligt nok, det er ikke så meget det; men så er det svært, så skal det opstå på 2 timer, så er rammerne ikke til at tænke videre. Så bliver sådan lidt, at så afvikler vi det her, fordi vi skal komme og give en tilbagemelding på et kursus, ja det gør vi hip hurra, og det skal vi nok gøre det og stå og sige nogle pæne ting, men det rykker ikke"	"… but anyway there are some colleagues who sit there out of bound duty, right. That is understandable, it is not as much that, but then it is hard, it has to occur within 2 hours, the setting is not meant for thinking any deeper. So it becomes a bit that we do it and report it during the course, yes, we do that hurrah, and we of course have to do it, go there and say some nice things, but it does not change a thing"

Karl sees opportunities in what QUEST presents but is also sceptical with regard to how much it really can change his and his colleagues' everyday teaching. Karl wants to assist his colleagues and furthermore expresses intentions of becoming a municipal consultant within science teaching. In the interviews, Karl accentuates his intention to help his colleagues in their science teaching as part of his career foreground.

9.3.4 *Laila, 51 Years Old, 21 Years as a Teacher: Participating Science Teacher from Daffodil Municipality*

Laila originally started a chemical engineering education but later changed to teacher education. She has taught children with special needs for many years and taught many different subjects in primary as well as lower secondary school. She has only been at her present school for a year, and here she teaches three science subjects: physics/chemistry, biology and geography. She would very much like to collaborate with colleagues on developing science education.

Danish interview transcript	English translation
"... for det første så giver det, tror jeg noget at være på kursus sammen det gør det altid. Men men netop med den der fordi det har været så hårdt at skulle stå for noget naturfagsudvikling på skolerne, hvis man ikke har opbakning, er det de færreste af os [der her haft]. Jeg har jo så været på hold med 2, der havde [opbakning], fordi de også var sammen, men begge to er så rejst fra skolen nu. Så men så alle vi andre det har været som at slå i en dyne, vi skiftede leder midt i det hele og så røg alle timerne og så forsvandt det hele bare bum"	"... first of all I think being at a course together is valuable, it always is. However, it has been hard to be responsible for science teaching development at the schools, if you do not have support, it is the fewest of us who have had [support], I have been on a team with two who had [support], because they were together, but they both are gone from the school now. Therefore, for us others it has been like hitting a cushion, we changed the leader halfway and then all those hours went and then it all disappeared - boom"

Danish interview transcript	English translation
"jamen foreløbig så forsøger vi at lave de der møder, man da vi samtidig skal lægges sammen med naboskolen, så er der rigtig rigtig rigtig mange møder lige nu. Og det er vi nødt til at tage højde for også, så man ikke drukner folk i et eller andet, så vi prøver at lave det som sådan noget foræringsnaturfag, for det man får som en foræring tager man som regel imod"	"well for the time being we try to arrange all these meetings, but as we are also being merged with the neighbour school, then there are very, very, very many meetings at the moment. This we have to take into account, that we do not drown people in something, so we try to make it a free takeaway science teaching, because what you get free you usually accept"

Laila had hoped that the QUEST programme would meet her previous experiences with another PDP. She wants to help her colleagues but finds that the ongoing merge with a neighbouring school makes it difficult for her. She would not mind leaving lower secondary teaching. Her foreground is to leave her present teaching position in order to change to teaching primary science and other subjects in primary school.

9.3.5 Diversity in Foregrounds

Relating the individual perspective from the analysis here in act 2 to the PDP's offered foreground gives us four different stories. It seems that the QUEST's anticipated foreground and the participating teachers' individual career foregrounds align in different ways. Based on the presented data and analysis, we constructed four different types of alignment: convergent, parallel, open ended and challenged. Poul's career foreground is *convergent* with QUEST's intentions of creating a strong local professional learning community. He finds that QUEST's collaborative agenda supports him and his colleagues in their daily work with planning and teaching. His career foreground is to continue teaching at the same school, which aligns very well with his experience of QUEST as being supportive. Birger's career foreground is *parallel* to QUEST's collaborative agenda. He hopes that QUEST will help improve local collaboration but already finds that he and his colleagues have a form of collaboration that works. He relates well to the QUEST agenda because it agrees with his view on science teaching. His career foreground is to start a career outside the school, which QUEST does not support. Karl's career foreground is *open ended* as he sees opportunities in QUEST but also limitations. He finds the intentions of QUEST promising but cannot see how they can be fulfilled within the given municipal framework. His career foreground is very explicit as he sees himself as a future municipal consultant, but he has not been able to make this career move yet. He is undecided about where his career will go. Laila's career foreground is *challenged* as she finds it very difficult to convince her colleagues about the development expected by the QUEST PDP. Her foreground is uncertain, as she has told her school leader in relation to the ongoing merge that she is willing to leave lower secondary teaching to become a primary school teacher. The different relations between individual and systemic perspectives are summarised in Table 9.1.

This diversity of the four presented teacher foregrounds raises the question of how or even whether a PDP will ever be able to meet each teacher's personal and individual foreground when participating in a PDP.

Table 9.1 Overview of relations between the intended PDP foreground and the individual career expectation foreground

Poul	Convergence:
Marigold municipality	Uses the provided learning opportunities in QUEST seamlessly for his personal and his colleagues' development
Birger	Parallel:
Daisy municipality	Acts autonomously to support structures, his personal development agenda more or less disregards the support from QUEST and other systemic supportive networks
Karl	Open ended:
Dandelion municipality	Addresses the gap between the systemic intentions in QUEST and his actual personal possibility to fulfil the intention. Karl is searching for stepping stones that can connect his personal foreground with the systemic supportive network offered by QUEST. He wants to keep all career options open both locally at the school and in relation to other opportunities within the municipal system
Laila	Challenged:
Daffodil municipality	Addresses the gap between the systemic intention of local collaboration and her actual personal possibility to fulfil the intention. Laila is considering leaving lower secondary teaching

9.4 Third Act: Discussion

When the teachers participating in QUEST are asked about their learning outcome using standard questionnaires as assessment tools, they express satisfaction with their participation in the QUEST rhythm. However, when they are asked outside the framework of QUEST about their experience with the relation between the PDP and their career consideration, other experiences are expressed. This accentuates the need for awareness of how to assess the outcome of a PDP. Is it appropriate to limit the assessment within the framework of the given PDP, or do you also have to focus on more lasting and deeper impacts of the PDP on the participating teachers' professional careers?

In this chapter, we set out to question whether PDPs that are designed to accommodate and improve individual teachers' professional capacity are more effective. The answer seems to be *it depends on how you assess the teachers' learning outcome!* In the following discussion, we will take up some of the aspects that our study has uncovered.

As mentioned earlier, different measures were taken to align the purposes of QUEST with the teachers' foregrounds:

- In advance of the modules, teachers were asked what concrete themes would be meaningful for them to work with.
- Aims and intended activities were outlined at the beginning of course modules.
- Teacher learning outcomes were assessed in open-ended questionnaires. Responses were used formatively to improve new activities.

- The QUEST rhythm was continuously used as a tool for dialogue with peers and instructors. The dialogue provided opportunities for assessing joys, frustrations, didactical reasoning, etc. leading to better insight into what works and what does not work for the teachers.

All these channels for alignment created a more aligned understanding between participating teachers, consultants, educators and instructors of the aims, options and limitations of QUEST as a PDP as seen in the focus group interview data. Still the individual teachers have their own view of what a PDP like QUEST can contribute to their personal teaching and development.

The different teacher career foregrounds found in the QUEST project can inform the wonderings of leaders and administrators reported by Talbert (2010). If leaders and administrators acknowledge that teachers' participation in PDP is guided by their own individual agendas, then perhaps they will not be surprised by the fact that teachers do not engage in PDPs as the planners intended. Furthermore, they may even be able to design PDPs to be more inclusive and participant sensitive.

Our findings accentuate the importance of awareness of the individual teacher's foreground in understanding their present commitment to the systemic expectation of the ongoing PDP. The individual foreground is part of a personal interpretative framework that guides the teachers' daily teaching and any change initiative affecting it (Daugbjerg 2015; Kelchtermans 2009). PDPs that create cohorts of colleagues to support the change and growth of the teachers within schools find that creating learning communities within schools supports change in teachers' beliefs, practices, and knowledge and that it helped to foster a more collegial learning community (Khourey-Bowers et al. 2005). Our inclusion of the personal level in the research on PDP shows that a condition for fostering a PLC is what the individual teacher sees as foreground for participation in the PDP and how his or her career foreground is aligned with this. Here it is important to recall that the notion of foreground encapsulates the teachers' present subjective perception of any given set of opportunities. This perception is liable to change when circumstances change and other aspirations, hopes and frustrations gain more importance.

If we return to the four teachers we interviewed in detail, we can see that two of them – Birger and Laila – in different ways expressed a wish to leave lower secondary science teaching. The QUEST PDP was intended to develop lower secondary science teaching. For this reason alone, QUEST did not coincide with their personal foreground. It is most evident in the way that Birger was not referring to QUEST activities in the interviews. Laila on the other hand tried but met organisational obstacles that made it very difficult for her to fulfil the intentions of the QUEST PDP. Karl also faced organisational obstacles but found relevance in the QUEST activities even though he could hardly implement them locally.

The task of making one PDP fit all participants is likely unsolvable. A solution could be better communication between the organisers of a given PDP and the future participants, as a way to align the participants' expectations with the PDP. Traditionally consultants and managers from schools, municipals, teacher educations, universities, etc. plan PDPs. Rarely the future educators on the courses or the participating teachers are involved. Obviously, this inclusion will complicate the planning as more opinions will be introduced, but this negotiating phase can adjust the expectations of all involved parts. Such planning with the participants will:

- Clarify the aims of all involved parts
- Clarify the opportunities offered to the participants
- Help educators to address the needs and expectations of the participants
- Help managers to state realistic expectations
- Clarify the limitations of all involved parts

Some of these elements of accommodation were included in QUEST as mentioned above, but still some of the participating teachers experienced difficulties in getting benefit from the PDP initiative. External factors such as a school merge influenced Laila's outcome of QUEST; such changes in the local agenda are unfortunate but will occur in larger PDPs. The case of Laila illustrates where a careful uncovering of her foreground would not have been able to counteract events interfering with the agenda of the PDP.

However justified a critique of the one-size-fits-all PDP approach can be, this present study indicates that the idea of a one-PDP-satisfies-all is likely a utopia. This should however not make PDP designers give up on the ambition to satisfy – nearly – all participants, but the important point is not to postulate that any given PDP will reflect that all experienced outcomes are equally successful. The application of two different approaches – systemic and individual – to PDP assessment furthermore has stressed that the success of a PDP is very dependent on the assessment perspective.

9.5 Conclusion

In this chapter, we set out to investigate whether PDPs that are designed to accommodate and improve individual teachers' professional capacity actually are more effective. Foreground is used as a notion to compare how a systemic designed foreground in a PDP matches individual perceived foreground by participating in the PDP. These two perspectives provide an intertwined analysis that leads to *three* major findings. *First*, the analysis of the presented teacher foregrounds indicated that teachers who meet some kind of kinship between on the one hand their own personal career foreground and on the other hand the systemic PDP agenda and organisation were more positive towards the PDP intervention. *Second*, four different foreground relations were found in our material: (1) increasingly convergent entangling foreground, (2) mutually enriching but distinctly parallel personal and systemic foreground, (3) open-ended foreground relation that keeps different career options open and (4) challenged career foreground that limits meeting the objectives of the PDP and makes the participant want to change teaching tasks. *Third*, teachers' foregrounds form a personal interpretative perspective through which the teachers relate to the local municipal framework and the PDP.

We set out to trouble whether PDPs that are designed to accommodate and improve individual teachers' professional capacity really are more effective. Our findings accentuate the relevance of designing PDPs so that the participants' expectations are brought forward and included in the design process. A way to do this can be to include an exercise in the start-up of PDPs that makes the participating teachers explicate their foreground and expectations. These foregrounds can then be aligned with the purposes, options and limitations of a given PDP. This would make it clear which of the teacher's individual career expectations the given PDP can meet and what part of the teacher's foreground must be pursued in another context.

Another important finding is also that assessment of PDPs can benefit from a systemic approach as well as an individual participant approach because the combination brings forward a richer description of the teachers' outcomes in the PDP.

References

Alrø, H., Skovsmose, O., & Valero, P. (2009). Inter-viewing foregrounds: Students' motives for learning in a multicultural setting. In M. Cesar & K. Kumpulainen (Eds.), *Social interactions in multicultural settings* (pp. 13–37). Rotterdam: Sense Publisher.

Berry, K. S. (2015). Research as bricolage. In K. Tobin & S. R. Steinberg (Eds.), *Doing educational research* (2nd ed., pp. 79–110). Rotterdam: Sense publishers.

Daugbjerg, P. S. (2015). Science teachers' foreground for continued professional development. *Journal of the European Teacher Education Network, 10*, 125–135.

Daugbjerg, P. S., & Sillasen, M. K. (2016). The relationship between science teachers' career expectations and different professional development experiences. *Journal of the European Teacher Education Network, 11*, 100–111.

Fullan, M. (2007). *The new meaning of educational change* (4th ed.). New York: Teachers College Press.

Gess-Newsome, J., & Lederman, G. L. (1999). *Examining pedagogical content knowledge*. Dordrecht: Kluwer Academic Publishers.

Hewson, P. W. (2007). Teacher professional development in science. In N. G. Lederman & S. K. Abell (Eds.), *Handbook of research in science education* (pp. 1177–1203). Mahwah: Lawrence Erlbaum.

Jackson, D., & Temperley, J. (2007). From professional learning community to networked learning community. In L. Stoll & K. S. Louis (Eds.), *Professional learning communities: Divergence, depth and dilemmas* (pp. 45–62). Maidenhead: Open University Press.

Kelchtermans, G. (2009). Career stories as gateways to understanding teacher development. In M. Bayer, U. Brinkkjær, S. Rolls, & H. Plauborg (Eds.), *Teachers' career trajectories and work lives: An anthology* (pp. 29–48). Dordrecht: Springer.

Khourey-Bowers, C., Dinko, R. L., & Hart, R. G. (2005). Influence of a shared leadership model in creating a school culture of inquiry and collegiality. *Journal of Research in Science Teaching, 42*(1), 3–24.

Luft, J. A., & Hewson, P. W. (2014). Research on teacher professional development programs in science. In N. G. Ledermann & S. Abell (Eds.), *Handbook of research in science education* (Vol. II, pp. 889–909). New York: Routledge.

Nielsen, B. L., & Sillasen, M. K. (2014). Science teachers' individual and social learning related to IBSE in a large-scale, long-term, collaborative TPD project. In C. P. Constantinou, N. Papadouris & A. Hadjigeogiou (eds.) *Science Education Research for Evidence-based Teaching and Coherence in Learning: Proceedings of the ESERA 2013 Conference. Strand 14*. Cyprus.

QUEST (2012). *En didaktik for QUEST [Trans.: A didactique for QUEST] http://q-model.dk/fileadmin/site_files/projects/quest/Filer/Materialer/En_didaktik_for_QUEST_2012.pdf. Accessed 10 Aug 2016.*

QUEST. (2015). *Qualifying in-service education of science teachers*. Retrieved from http://quest-projekt.dk/. Accessed 10 Aug 2016.

Rolls, S., & Plauborg, H. (2009). Teacher career trajectories: An examination of research. In M. Bayer, U. Brinkkjær, S. Rolls, & H. Plauborg (Eds.), *Teachers' career trajectories and work lives: An anthology* (pp. 9–28). Dordrecht: Springer.

Shulman, L. (1986). Those who understand: Knowledge growth in teaching. *Educational Researcher, 15*(2), 4–14.

Sillasen, M. K., & Valero, P. (2013). Municipal consultants' participation in building networks to support science teachers' work. *Cultural Studies of Science Education, 8*(3), 595–618.

Skovsmose, O. (1994). *Towards a philosophy of critical mathematics education*. Dordrecht: Kluwer Academic Publishers.

Steinberg, R. S. (2015). Proposing a multiplicity of meanings. In K. Tobin & R. S. Steinberg (Eds.), *Doing educational research* (2nd ed., pp. 111–132). Rotterdam: Sense publishers.

Talbert, J. E. (2010). Professional learning communities at the crossroads: How systems hinder or engender change. In A. Hargreaves, A. Lieberman, M. Fullan, & D. Hopkins (Eds.), *Second international handbook of educational change* (pp. 555–571). Berlin: Springer.

van Driel, J. H., Meirink, J. A., van Veen, K., & Zwart, R. C. (2012). Current trends and missing links in studies on teacher professional development in science education: A review of design features and quality of research. *Studies in Science Education, 48*(2), 129–160.

Webb, T. (2009). *Teacher assemblage*. Dordrecht: Sense Publishers.

Chapter 10
Enacting Citizenship in Ordinary School Science Through Deliberative Communication

Gerd Johansen, Guðrún Jónsdóttir, and Stein Dankert Kolstø

10.1 Introduction

We are in a science lab in a Norwegian upper secondary school 2 months before the end of term. The students are aged 16, and this is their last year of compulsory general science. The students are divided into groups of four or five, as they are to work together this lesson. There are several practical activities in electrochemistry on the schedule this day. The students are to make a tentative explanation and present it after each activity. The teacher gives the groups two task sheets to share, and then the teacher's voice sounds loud and clear: "Put away all loose items on your desk". The teacher then talks to the class about the equipment they are to use and finishes with: "Then, you should all be able to collect the equipment without making noise and using up too much time. And all of you must have lab coats and protective glasses, because when we use some of the solutions we need glasses". There is a question from a student if they are going to report the activities directly onto the task sheet. The teacher replies: "Yes, you can do that, and you discuss […] It is smart to take pictures, because I think that you might possibly be chosen for the final oral examination. Anything else?" – there is no response. The groups start working.

Does this have anything to do with citizenship?

The above is a typical introduction to practical work in science. However, practical work as such is not what we focus on in this chapter. Nor do we focus on learning outcomes in the traditional sense. Rather, we explore the idea that citizenship education might take place within the context of teaching and learning established

G. Johansen (✉) • G. Jónsdóttir
Norwegian University of Life Sciences, Ås, Norway
e-mail: gerd.johansen@nmbu.no

S.D. Kolstø
University of Bergen, Bergen, Norway

K. Otrel-Cass et al. (eds.), *Cultural, Social, and Political Perspectives in Science Education*, Cultural Studies of Science Education 15,
DOI 10.1007/978-3-319-61191-4_10

science content, in other words, within ordinary science education. This perspective is not currently being an emphasised focus of empirical research. Specifically, we ask whether patterns of deliberate communication might be involved in students' learning processes in science and thus involve aspects of, and preparation for, citizenship. While students are discussing how to do, what they observe and how they want to explain the practical activities, they might have to deal with disagreement, differing interests and making decisions where they cannot foresee the outcome (even if the teacher can), and perhaps they might question the authority of the teacher or subject matter. In this kind of communication, which Tomas Englund (2006) terms deliberative communication, we interpret the actions of students as citizens.

Under the entry science for citizenship in the *Encyclopedia of Science Education*, Virginie Albe (2015) outlines two main views on citizenship within discussions about science education. Firstly, it is important for the populace to appreciate and see the beneficial role of science and technology in bridging the gap between science and society. Secondly, the "more commonly held view is that 'science for citizenship' invites science educators to engage in a deep reformulation of a school science curriculum that no longer meets the needs, interests and aspirations of young citizens" (Albe 2015, p. 905). This second view is often related to socio-scientific issues; see, for instance, Stein Dankert Kolstø (2000) and Daniel Birmingham and Angela Calabrese Barton (2014) for an activist approach to such issues. However, in this chapter, we adopt neither of these views; rather, we explore the possibilities for students dealing with the established school science knowledge to practise citizenship. Discussions in science education research on the teaching of socio-scientific issues typically focus on involving students in such issues. However, deliberate communication is not mentioned, e.g. in the *Second International Handbook of Science Education* (Fraser et al. 2011). Furthermore, in a review on discourse practices, Gregory Kelly (2014) highlights research issues such as argumentation and participation in discourse practices in order to learn science and its epistemology and also to develop students' identities, access to practices and equity. The notion of discourse and the communication of established science do not seem to be directly coupled with citizenship education. This decoupling could of course be due to the notion that citizenship education is not researchable within the context of ordinary school science. However, as this might be wrong, we seek to trouble this decoupling by exploring citizenship education *within* content-focused science education, focusing on the concept of deliberative communication.

In the first part of the chapter, we outline our position on citizenship and science education, before continuing to the analytical framework, deliberative communication. This framework will be operationalised at the start of each empirical section. The examples provided in the empirical sections stem from extensive video material from two science classes. In the final section, we conclude that instances of deliberative communication were present in the science communication explored, indicating that in ordinary science classrooms deliberative communication is possible to identify – and thus make researchable. Moreover, we discuss whether deliberative communication might overlap with professional scientific practices; if they share virtues, it might become feasible to combine these in ordinary science teaching and learning.

10.2 Science Education, Citizenship and Deliberative Communication

Before the analytical framework – deliberative communication – is presented, the role of science education in educating for democratic citizenship needs to be unfolded. Terms such as "democracy", "citizenship" and even "science education" are slippery and can be explicated in several ways. Moreover, discussions on education for citizenship generally include several disciplinary areas and thus become complex to handle. Ralph Levinson (2010) argues that ubiquitous rhetoric might disguise real differences and as an extreme consequence lead to disempowerment and anti-democratic practices. Douglas Roberts and Rodger Bybee (Roberts and Bybee 2014) point out that most often science for citizenship is connected to socio-scientific issues or science, technology and environmental issues in society. Ian Davies (2004) states that when the learning of science is linked to citizenship, it is required to recognise what science is all about, in other words, a firm link to "nature of science". Moreover, Davies raises concern over simplified notions of citizenship education within science education research.

There are many and varied definitions and usages of the term democracy (Crick 2002). We draw on Bernard Crick (2002) and approach democracy as a way of acting towards others – the enactment of citizenship. This enactment involves participation in egalitarian activities and respecting other participants' rights (Crick 2002). Terence McLaughlin (1992) argues that the concept of citizenship is complex and contested, as views vary greatly on, for example, individual agency: what is required to involve in (what) political processes and social prerequisites necessary for citizenship. Based on previous research in the field of science education, Levinson (2010) proposes four frameworks for dealing with democracy in science education: deficit, deliberative, science education as praxis, and science education for conflict and dissent. Each of these positions the citizen differently. Levinson raises a critique against the framework of deliberative democracy, questioning citizens' possibilities for participation in democratic dialogues.

As the possibility of citizens' participation is an open question, we see it as feasible to explore how a deliberative approach is realised in ordinary school science. The starting point is that students are citizens here and now, as well as are preparing for future participation in a democratic society:

> We must take the child as a member of society in the broadest sense, and demand for and from the schools whatever is necessary to enable the child intelligently to recognize all his [sic!] social relations and take his part in sustaining them. (Dewey 2015/1909, p. 230)

Participation in democratic dialogues might thus include ordinary school practices.

10.3 Deliberative Communication

Deliberative communication is one of several approaches for enhancing citizenship in schools. Tomas Englund's (2006) framework of deliberative communication builds on and takes its inspiration from the pragmatic tradition as well as deliberative democracy theory. In the pragmatic tradition, communication is vital for participation in society. In *Democracy and Education*, the pragmatist John Dewey (2007/1916) states that the school is important to laying the foundations for developing students' deliberative capacities. This is considered a long-term process.

Jürgen Habermas is considered the foremost spokesperson for the deliberative democracy theory. In deliberative democracy theory, social integration is linked to collective communicative processes of "knowledge and will-formation with regard to what is normatively valid and 'true'" (Englund 2006, p. 509). The ideal is of a public discourse where equal participants obey "the unforced force of the better argument reaching (or approximating) consensus along this road" (Kock 2007, p. 181). A criticism of this rationalistic approach is that it potentially allows those most skilled in rhetoric to control the decision in their own favour (Rienstra and Hook 2006). However, this does not imply that arguments play no role in peoples' will-formations and deliberations, e.g. in democratic processes. Our position is therefore that practices involved in deliberate communication are important to include in education.

According to Englund (2006, 2015), deliberative communication implies communication in which:

- Different views are challenged to make space for different arguments.
- Tolerance and respect amongst the participants, respect for the concrete other and participants learning to listen to the other person's argument.
- There is an element of collective will-formation, an endeavour to reach consensus or to form a conclusion or, even, to agree on disagreeing.
- Traditional perceptions and authorities can be questioned.
- There is opportunity for students to communicate and deliberate without teacher control.

The first three are considered the inner core where different views are presented, where the concrete other is treated with respect and there is a collective will to reach consensus – or a procedure for consensus (Englund 2006). In specific, Englund (2006) states that:

> Deliberative communication (used in an appropriate way) can also contribute to meaning-creation and knowledge-formation among students in most subject areas, even within traditional school subjects and areas that seem to be a long way from subjects closer to the democratic foundation of schools. (Englund 2006, p. 505)

This statement explicates the position we are exploring. Using Englund's five categories as an analytic tool, we seek to explore the students' space for deliberative communication – and how they make use of it – while they participate in ordinary school science.

10.4 Context and Method

The Norwegian science curriculum document states the significance of practicing and fostering active and responsible citizenship (Utdanningsdirektoratet [Norwegian Directorate for Education and Training] 2013). Most school science content is established undisputed knowledge, not usually thought of as involved directly in democratic processes. Our point of departure is the notion that there might be possibilities for deliberative communication also in ordinary teaching-learning situations.

Practical activities are apt when it comes to studying phenomena which rely on verbal communication. Students talk to make decisions and assessments during practical work. However, we agree with Bent Flyvbjerg (2001) that power relations cannot be ignored in communicative situations when searching for "the better argument". However, in this chapter, we choose not to focus on such power relations per se, although we acknowledge that in the school science practice, the established body of knowledge as well as school organisation through, e.g. exams and markings, exercises power.

In this chapter, we refer to two practical activities that are very common in compulsory Norwegian school science when dealing with the topic of electrochemistry, namely, electrolysis of copper chloride and burning magnesium. The students (aged 16) are given a task sheet (recipe) which provides a list of equipment, outlines a procedure and has blank spaces after the words "observations" and "tentative explanation". Students' own formulation of observations and tentative explanations were meant to ensure that students had to talk, assess and deliberate with each other – and not simply look up the right answers on the Internet or in their textbooks. The teacher did not provide the students with an explicit purpose for this; however, observations and explanations are common features of practical activities in science. Moreover, the teacher did not provide the students with any support on how to conduct a deliberative conversation. In other words, we are exploring naturally occurring communication. These lessons took place in April, so it is assumed that students and teachers are fairly well acquainted and none seem to be troubled by being recorded. Thus, we assume that students are behaving as they usually would in the dialogues.

The teaching-learning situation was videotaped with one whole-class camera, and when students start to engage with the practical activity, one student from each group is fitted with a portable camera. Those who did not want to be recorded are in a separate group according to principles of volition (Thomas 2011). The third author was a participant observer in the classroom while collecting the video material. The first and third author transcribed the material. Video material provides the researchers with the possibility of (re)viewing and discussing the material according to theoretical frameworks (Goldman 2007). We have on this basis chosen excerpts suitable for discussing the presence of deliberative communication. Our interpretations, based on Englund's characteristics, were discussed amongst the authors. Through these discussions, we operationalised Englund's framework to enable consistent analysis of observed practices. Although specific features of each of Englund's categories were identified, we do not claim that these features are exhaustive.

The transcripts that are presented here will never give full justice to classroom activity. There are often two or three parallel conversations within a group, and there will often be activities not captured on camera, so in our presentation of communication, we have reduced complexity to facilitate accessibility.

10.5 Deliberative Communication During Practical Work

Although we analyse communication during learning activities, we are not attempting to provide a comprehensive analysis of deliberative communication. Rather, we want to provide some examples of students' talk that can be interpreted as deliberative – or not – according to the characteristics in Englund's framework. The characteristics will be explicated prior to the interpretation of the excerpts. Moreover, we do not regard these characteristics as totally disjointed. The examples we provide will form the backdrop for the oncoming discussion.

10.5.1 Challenging Views and Making Space for Different Arguments

Concerning the analysis of this first of Englund's characteristics, we do not focus on the science content, e.g. how "correct" the statement is or what it means for the students' "understanding". Rather, we look for possibilities to state different views and challenge these. When dealing with practical work and conceptual understanding, it is possible to challenge views and arguments even if the focus is on established science (Engle and Conant 2002). Observations are not necessarily obvious for students (Bergqvist and Säljö 1994); however, when observations are agreed upon, they are regarded as facts by the discussants (Hamza and Wickman 2013). When students are to "make tentative explanations", they have to interpret observations and connect these to theoretical facts and principles. Some explanations will be better than others as they are more aligned with the norms of the field (Knain 2005).

During their process of learning, the students might try out arguments on each other. They express themselves with varying degrees of certainty in statements and questions. Challenging ideas or proposals can be done directly by open disagreement or more indirectly by asking questions or proposing alternatives. We make a distinction between challenging an idea and challenging the person who proposed it. In other words, it is possible to challenge each other's and other person's ideas. Participants might make space for different views by listening to each other and following up others' ideas or proposals, or signalling disinterest.

A group of five students carry out "copperplate coin" – electrolysis of $CuCl$. This excerpt shows how difficult it is to observe and to infer from observations. Ellen raises her eyes and looks directly at the teacher who is passing their desk.

Excerpt 1

1	Ellen:	It starts to etch, it etches, why?
2	Teacher:	What do you [Ellen] mean by etch?
3	Ellen:	It falls off (she taps the electrode onto the bottom of the beaker)
4	Jon:	It does not etch
5	David:	It does not etch because it
6	Jon:	Because it is just the coating that falls off
7	Teacher:	There are two different ideas that are proposed, is it something outside or is it something part of the coin that falls off?
8	Ellen:	How long do I have to sit like this?

Ellen's question (directed to teacher?) opens up for an idea not previously discussed in the group. Upon the teacher's request, she explicates her idea somewhat (line 3). The two boys sitting next to her, Jon and David, openly disagree ("it does not"). However, they do not explicate why they inferred differently (i.e. answer the question of why the coating falls off). This becomes a disagreement that is just partly unpacked. It can be seen as two sides challenging each other. Here the teacher does not take a stance as to whether it is Ellen or Jon and David that have the "right" idea (line 7). The teacher's cue to unpack the argument, where the teacher sums up the two different views (line 7) – to allow for the students' arguments – is not followed up; rather, Ellen chooses to change the subject. It is possible that she expected the traditional yes/no answer, and the teacher's answer dissuaded her from following up. Or perhaps, Ellen chooses to drop the issue because two of the other students disagree with her. In ordinary school science, a prevalent norm is that it is important to arrive at the right conclusions.

Another group (four students) is doing the same experiment, and they are trying to explain what is going on. As a battery is part of the equipment, surely it has to do something with the explanation?

Excerpt 2

1	Sonja	What is actually going to happen?
2	Liv:	Eh, now we are making, yes, no, I don't think we're charging the battery or anything like that
3	Tom:	Yes?
4	Tom:	Or, we're charging that rubbish there (points at equipment/beaker)

5	Liv:	Yes, one of those things is charged
6	Tom:	What's most logical?
7	Liv:	Well, the battery already has electrons
8	Chris:	How do you charge water?
9	Liv	No, but it's possible. But it is inside a circle. It goes in a circle. That's what it does. It comes…
10	Tom	But look, it goes the wrong way a bit. So it goes out of the battery. (continues explanation)

Sonja's question seems to trigger Liv to start an explanation. Liv and Tom are then trying out arguments on each other, in a rather informal manner, for instance, the informal "rubbish" in line 3. By using utterances like "eh", "yes" and "no" in succession, Liv in line 2 signals uncertainty or openness in addition to an idea (it is *not* charging the battery). This uncertainty/openness probably allows Tom to engage in the conversation. He seems quite hesitant in the beginning. In line 5, Liv is outlining two alternatives and thus challenges her own initial idea. By doing so, it is perhaps easier for the other students to challenge her idea as well. Tom follows up (line 6) and challenges these views by asking for a connection between the two alternatives and "what is most logical". All the students in the group seem to listen to each other, and they partake in making different views visible. However, Chris's question (line 8) has no great impact on the direction of the conversation, perhaps because this is by then an "old" idea for Liv and Tom – they have moved on to the idea of an electrical circuit.

The students in excerpt 1 do not follow up each other's different ideas by making space for alternative arguments or asking questions of one another. There seems to be a somewhat direct form of challenging an idea between the students. This excerpt does not satisfy the criteria for a deliberative approach for how to challenge views and to allow for different views. In excerpt 2, there is a more deliberative approach, as there are more utterances signalling space for more views and for challenging these by putting forwards questions and alternative views. However, in both groups, each student's voice seems to have a different impact and weight. The analysis of these two excerpts indicates that in students' discussion in ordinary science education, it is possible to identify the presence, and absence, of practices reflecting Englund's first category: challenging views and making space for different arguments.

10.5.2 Tolerance and Respect

Tolerance and respect are elusive concepts. The concept of tolerance can be explicated from diverse theoretical stances and is thus hard to operationalise (Afdal 2010). Moreover, it is beyond the scope of this chapter to partake in the debate on the distinction between these two (neighbouring) concepts. We regard the presence or absence of tolerance and respect as parts of the communication on subject matter. Respect, as we see it, is when something or someone is given importance and thus held in high esteem, and tolerance is about how differences are handled. Tolerance makes it possible to disagree strongly and at the same time understand and take into consideration the other person's line of argument. The acts of tolerance might be active or passive. Active tolerance can be actions such as asking questions even when in disagreement as well as trying to explicate one's own standpoint. The contributions of others are treated as valuable. A more passive form of tolerance in a classroom setting can be to let the other "carry on" without interference. Lacking tolerance is observable in actions such as denying access (to things and conversations). Lack of respect (disrespect) can within a classroom setting be subtle, such as not letting someone finish an utterance, yawning, turning one's back to the other person and smiling in an overbearing manner. More direct or active disrespect can be connected to, for example, verbal abuse.

Moreover, the expressions of tolerance and respect are culturally dependent. Expressions of respect amongst 16-year-olds might not be identical to the views educational researcher's hold of such expressions. However, expressions of tolerance and respect, or lack thereof, are approached by both inspecting verbal and non-verbal communication. It is more difficult to make interpretations based on non-verbal communication, mostly because the actions are subtle. We have watched the video together with the transcript in order to make interpretations well argued for.

If we return to excerpt 1, Ellen is looking intently into the small beaker. She has just interrupted Jon, while he was trying to explain electrolysis to the other students. Clearly, she regards that her observation (something is falling off the electrode that she infers etching) is important – and she poses a question (line 1). The two boys, sitting left and right, disagree with her inference (line 4–6). It seems that Ellen is not interested in the alternative idea (line 8). The two other students in the group are not contributing to this passage: The other girl takes out her mobile phone to take a picture. This situation can be interpreted as the students respecting each other and showing tolerance, passively, by letting the other carry on and not enlarging the disagreement. However, this situation might also be interpreted as a kind of passive disrespect. Ellen is not interested in Jon and David's ideas about the phenomenon; she interrupts Jon and turns towards the teacher. Jon and David are also not very interested on how Ellen arrives at her conclusion. The two students on the other side of the table are not especially interested in any of this. Perhaps the students' way of handling the situation is due to the need to "get the job done", and that means that not all contributions can be explored at length. So even if the students are "harmoniously working side by side", it can be interpreted as them mutually ostracising each other.

In excerpt 2, all the students have their attention on the equipment, well not totally. When the excerpt starts, there is a parallel conversation "Does anyone have a pencil?" This topic (pencil) is lost when they together start to generate and challenge ideas. Tom is eagerly participating – his speech is fast and at a rather high volume, but he sounds friendly. In lines 9–10, he is interrupting Liv so he can say his piece. However, none seem to be upset by this, and they continue to explore ideas. By asking a direct question of the other students (mostly to Liv?), Tom, in line 6, opens up the floor for the other students' contributions. As he stops talking himself, he signals to the other students that he wants their contributions. We interpret this as an act of tolerance. He could have continued in his rather boisterous manner and only been interested in his own ideas. Liv follows up Tom's question. On the other hand, Chris' question, line 8, is met with a kind of passive disrespect – it is ignored – even if the question of "how do you charge water" is a relevant and a challenging question in the given circumstances. Why are these two questions dealt with so differently? Perhaps it has to do with Chris's academic status within the class or that the other students regard his question as a side-track. Time will undoubtedly influence how they follow up ideas.

There are quite a lot of examples of what we interpret as passive disrespect in the empirical material. This can, for instance, be that a student is yawning when a peer is talking – or the teacher is talking. Or, it can be as shown in these two examples that some students' contributions are ignored. We have not however explored if these cases of ignoration are systematic. Thus, what seems like a friendly setting might actually exclude some from the possibility of contributing to the joint decisions. Although the operationalisation of this category needs further exploration, these analyses show the possibility of identifying empirically, at least partly, some of its characteristics.

10.5.3 Collective Will-Formation: Decisions

When collaborating on everyday activities, decisions will sometimes be explicitly agreed upon but often be an implicit part of the communication. In a group, someone has to take the initiative and responsibility to generate ideas and proposals that are laid out for decisions. According to Jane Mansbridge, Janette Hartz-Karp, Matthew Amengual and John Gastil (2006), "good atmosphere" and progress are important for group members during deliberation. The good atmosphere might be connected to the possibility to scrutinise proposals and ideas, i.e. closely related to respect and tolerance. Making progress (also) means to leave some issues behind when they are resolved, for the time being. This might, for instance, be seen when an idea that starts out as contested ends up as an implicit assumption in the continued discussion. Explicit decisions might be a proposal that is affirmed through the use of utterances such as "yes" or "ok" or a non-verbal nod. Obviously, implicit decisions are part of "carry-on modus" and might be hard to detect. Finding reliable ways to identify "traces" of implicit decisions would then be vital but is outside the scope of this chapter. Moreover, regarding implicit decisions, we encounter the problem of the demarcation of a decision – i.e. how "big" a decision must be to count as a decision. Also in need of further investigation is to develop ways to analyse to what extent the decisions are made collectively, for instance, if all students are agreeing or if some are more or less explicitly "delegated" the power of decision making – or if some decisions are just happening.

We return to the group in excerpt 2 a minute earlier. The teacher has stopped by and helped the students assemble the equipment before she leaves the group. The electrical cords are connected to electrolysis rods and to the battery. But, what about the bluish liquid in the small beaker?

Excerpt 3

1	Tom	But I thought we should put it all into that one (points at small beaker)
2	Liv:	Yes, but, we have to pour this into (big beaker)
3	Tom:	But [Liv], we could just as well put this straight into that (i.e. electrolysis rods into small beaker)
4	Sonja:	No
5	Tom:	It is large enough
6	Liv:	But, why should we to do that?
7	Tom:	But, shouldn't you not take this one (electric cord) to this (electrode)?
8	Sonja:	No
9	Liv	She [the teacher] said that we should have this one on that
10	Tom	Ok
11	Sonja	Ok, then we pour it into

In the beginning of this excerpt, there is clearly a disagreement on "how to do". Sonja's two loud and clear "no's" indicate as much (line 4 and 8). It is therefore interesting to explore the students' process towards making a decision. Tom argues against Liv in line 3 but does not produce a counterargument to Liv's question in line 6. He chooses to shift the topic slightly (line 7). Tom is problematising yet another aspect of the equipment set-up. In none of these issues do the two girls choose to support Tom's view, he loses the "battle". The need to make progress probably makes it necessary to make a decision, and Tom affirms "ok". However, this is a small decision and it will not make any difference – and they probably know that. So why do they make an "argument" out of it? Perhaps this is a way of discreetly showing who's in charge here.

When Tom agrees through his use of "ok", this can be understood as if the students made the decision collectively, even if Chris did not partake in the argumentation. He does not seem to have a strong opinion on this matter. Whether the best argument "won" is another question. Interestingly, some minutes later, the teacher stops by their desk and asks why they used the big beaker. Tom answers "I said the same thing, but they insisted". The "collective decision" is now pinned down on the girls. It can be interpreted as Tom distancing himself from the decision and, as a feature of the power distribution in the school system, where students will try to "look good" in the eyes of the teacher.

The task given to the students was to formulate a tentative explanation. What counts as an explanation might differ between different situations. However, there are established traditions for how scientific phenomena should be explained in school science. This limits the students' space for will-formation. While their starting points might be diverse, they are to arrive at a fairly predetermined destination – "the correct answer". In addition, the subject matter – electrolysis in this case – is difficult to grasp. We choose therefore to illustrate this point with a conversation where the teacher is also (partly) present. The teacher spends quite a lot of time together with this group. First, she leads the students through the observations, the smell of chloride and the coin's new colour – or "copperplate" as Linda puts it. Two minutes later, the teacher returns and asks whether the students have started to explain.

Excerpt 4

1	Charles	(talks slowly) It is because the copper sulphate the electrons come from
2	Teacher	(about to leave) You have to ask your fellow students if they agree
3	Charles	The electrons move because of. (To teacher:) This has nothing to do with the activity series, right?
4	Teacher	(turns toward Charles) There is no need to think about the activity series to explain this

Charles seems to want to try out his idea on the teacher as he talks more slowly than usual. In line 2, the teacher wants the students to try out this idea and thus to have a joint will-formation. For the student(s), it is also important to park ideas not relevant to think about, i.e. the activity series (line 3), in order to make progress. The "activity series" is resolved. However, the teacher stays with the students for half a minute longer, and Charles elaborates on why not to include the activity series in the explanation. The only girl in the group, Linda, tries to insert her explanation (of the electrolysis?) "Can it be because", and the rest of her meaning trails off as no one seems to listen to her. A while later, she has an idea which the teacher acknowledges and asks the students to investigate, but here and now she is sidetracked. Tim and Oscar do not contribute in this passage. When the teacher leaves their desks, Charles asks the other students "did you understand that?", and Tim answers yes. The students starts to talk – but not about the explanation. Then Charles exclaims.

Excerpt 5

1	Charles	Help me here! Ions from the liquid jump in a way, or electrons…
2	Linda	Can I have that one back (takes the carbon electrode from Charles)
3	Oscar	What are you thinking about, Charles? (yawns)
4	Charles	If you think [in terms of] reaction equation or
5	Tim	This one (points at the lab sheet – And the reaction equation for another experiment)

In line 2, Linda chooses to continue to carry out the practical activity: Why? Perhaps she expects that there will be more observations or perhaps it is a way to seemingly keep busy. Oscar does not say much during this lesson. He is, however, paying attention. All the students are paying attention. Charles is given or takes much of the responsibility to provide ideas for the explanation (e.g. line 1). When the teacher is present, Charles and the teacher talk. He tries out formulations on the teacher in the period between excerpts 4 and 5. After this, Tim starts to write their "joint" explanation – with substantial help from Charles. Linda and Oscar start to tidy up and leave the desk.

In this group, there is no doubt that the collective will-formation stalls – it is not collective. Perhaps the students are not sure of the purpose of this explanation, the required level of accuracy and how to connect the different observations and theoretical facts probably involved in the explanation. This, combined with the fact that there is *an answer*, might result in withdrawal from the collective responsibility. Moreover, the students are not enticed to partake and there are no sanctions. The collective report makes the individual contribution invisible.

Making collective decisions is difficult when one cannot foresee the consequences. This is perhaps the reason why some parts of descriptions and explanations were left as partly solved – before the students returned to the issue when writing the report. Moreover, to reach agreement – or disagreement – in science might be seen as requiring a certain cleverness, which discourages many students from partaking in the collective will-formation. However, from a research perspective, we were able to identify empirically collective will-formation during the students' discussions. Nevertheless, there is a need to explore further the characteristics of will-formation: how it is shaped through dealing with established scientific content and how it is linked to tolerance and respect.

10.5.4 Traditional Perceptions and Authorities Can Be Questioned

This characteristic is challenging, seen from a school science point of view. The subject matter the students are working with is very well established – it "works" and thus has authority in itself. To question the knowledge base of electrochemistry is more than one can expect from young science students. However, it is nevertheless possible to question or even critique authoritative sources (textbook, task sheet or teacher) as part of meaning-making. To have authority might mean to make the standards of subject matter visible (Engle 2011). This characteristic also deals with situations where students ask for the purpose of an activity or the relevance of the subject matter. When students ask: "why learn this?", it can be understood as a (constructively) critical approach to subject matter.

Excerpt 1 above is quite interesting from an analytical point of view. The teacher is not providing the answer Ellen asks for (line 1). Here the teacher falls out of the role as "the provider of answers" and instead opens up for the students to explicate their views (line 7). The teacher does the same "trick" as an answer to Ellen's question in line 8 ("How long do I have to…"). Ellen says in a tone almost playful: "You are the teacher here, so you have to know". In this utterance, Ellen is stating explicit expectations of the teacher. A teacher is to answer and thus has the right to decide what is good enough. This can be understood as handing authority over to the teacher – i.e. not to question authority. In excerpt 3, the students are also explicitly calling upon the authoritative sources of scientific knowledge – in this case represented by the teacher. In excerpt 3, line 9, when Liv says that "teacher said" it is an answer that Tom does not question. The teacher knows. On the other hand, the teacher's actions are not above questioning or critical comments. One example is when Tom has gone to collect some equipment. He and the teacher stand by the equipment trolley. They have the following conversation:

Excerpt 6

1	Tom:	We just need the one beaker?
2	Teacher:	Yes. (points at beaker.) it is over there, already filled with [copper chloride]
3	Tom	Aha!
4	Teacher	And then, [you need] a coin and a battery
5	Tom	But the others have already taken such a blue one (referring to battery)
6	Teacher	Yes, but perhaps your group ought to get yourselves organised?
7	Tom	Yes, we have! Everyone has taken what they need
8	Teacher	That's good!
9	Tom	It is you that's con…
10	Teacher	Yes, but you don't need to
11	Tom	Confusing me, you know

The tone of their voices is quite amicable. It is not a teacher-student conflict. It seems like Tom finds the teacher's mild criticism (line 6) unfair. The group had collected the necessary equipment! Instead of criticising the teacher for being unfair, Tom's statements in line 9 and 11 can be seen as mild critiques, where he takes part of the blame – for letting himself be confused. It seems that in this class, it is allowed to question the teacher and to make subtle critiques of the teacher, within limits – even if their purpose of it is to get the teacher to recognise her authority and to provide answers and standards.

This day the students also burned magnesium. The groups took turns by the ventilation cupboard. We return to the group who struggled to make an explanation in excerpts 4 and 5. Here follows an account of what they experienced – which was to them inexplicable. They take out a magnesium stick from the packet, and they try to set it on fire by using a candle. Nothing happens for one and a half minute. The researcher stops by and watches the students.

Excerpt 7

1	Tim:	Does it take this long?
2	Researcher	That stick, what is it made of?
3	Charles	This one?
4	Researcher	Yes
5	Charles	Magnesium
6	Researcher	Are you sure?
7	Charles	Dead sure
8	Researcher	It does not look like that to me
9	Charles	It says magnesium sticks (on the box label)
10	Linda	What happens is that it becomes black

Except for a layer of soot – or as Linda said “it becomes black” (line 10), there was no (chemical) reaction. The researcher questions the students’ assumption that the stick was magnesium (line 2 and 6), and in line 8 the researcher provides an argument for his question “it does not look like”. The students are here provided with a conflict between two authoritative sources: the researcher and the equipment in a “magnesium stick” packet laid out by the teacher. Two minutes later, the students try to burn a magnesium ribbon. It burns with an intensive white flame and Linda exclaims “wicked”. However, the students do not abandon their attempt to burn the magnesium stick. They try with several matches at once to increase heat, and the teacher brings a Bunsen burner. A couple of minutes later, the teacher returns and asks “Is it ok now?”. When it becomes clear to the teacher that it indeed did not work, she asks the students to give it up and move on to the copperplate coin activity. By then, the students have spent 10 min trying to ignite the magnesium stick. Later, when the teacher stops by this group’s desk and they are making an attempted explanation, the following conversation takes place.

Excerpt 8

1	Charles	Observation. What happened?
2	Teacher	Yes, what happened?
3	Charles	The magnesium ribbon started to burn, but that one
4	Teacher	Include rather, just include that the stick did not work
5	Charles	Shall we include that?
6	Teacher	You can very well do that. That it was very difficult, and the reason for it is difficult to give
7	Charles and Tim	Source of error
8	Teacher	Yes

This situation is not very interesting, seen from a scientific point of view. The students tried to burn magnesium oxide (MgO) – naturally unsuccessful. (The rod was supposed to be used in another type of flame tests and had a misleading label.) What is far more interesting from a “scientific attitude” point of view is that neither the students nor the teacher questioned the alleged magnesium stick. They did not even question it when they saw how easily the magnesium ribbon burnt – or when the researcher pointed out that the stick "did not look right" – it did not have a metallic colour. Why did the teacher not react? Obviously not because she did not know the difference between magnesium and magnesium oxide. Rather than trying to question and to find a rational explanation, they accept the vague "source of error" as a reason for the non-burning magnesium stick. A magnesium stick has to be pure magnesium, does it not? It is not to be questioned – nor is it to be deduced from the fact (stick's colour and how easily the pure magnesium ribbon burned) that this was not metallic magnesium.

Excerpt 7 shows how a student expresses opposition to the researcher's critique. Excerpt 8, in contrast, shows how students are not willing to question the combined claim made by the teacher and the label on a box with scientific equipment stating the rod was made of magnesium – even if evidence to the contrary was evident. In other words, from a research position, it seems possible even in such content-driven activities to find traces of questioning authorities.

10.5.5 Students' Opportunities for Deliberate Communication Without Teacher's Control

When doing practical work and other group activities, students have many opportunities to talk together without teacher control. One may nevertheless analyse the extent to which the teachers keep some sort of control over students' communication and the strategies used. Moreover, the way students use opportunities for deliberate communication will also depend on their ability to practise such communication. It then becomes relevant to look for instances where students somehow are taught or modelled deliberate communication.

During the introduction, the teacher several times states that students are to observe and make tentative explanations. She does not state explicitly that this is to be a collective group effort. However, she does so implicitly, by stating that all groups are to write their explanations on the board. Moreover, there is not, in the introduction before students start their work, any mention of how to talk together. For instance, the teacher does not give cues such as listen attentively, elaborate on ideas or to problematise content.

During the activities, the teacher walks around and helps the students. Some of the issues she helps with are of a factual nature – e.g. the question in excerpt 3 (lines 7–9) which deals with how to assemble the equipment. The equipment was not meant to be the issue of "thinking" – so by directing the students on this issue, they can move on to the task (observe and explain). In excerpt 4, the teacher directs most of the attention to one of the students when trying to establish a tentative explanation. In this material, there are few if any examples where the teacher explicitly includes and challenges all the students in the conversation and thus models argumentation and joint will-formation. On the other hand, it seems like the teacher chooses an approach that requires students' volition; see, e.g. excerpt 4 line 2.

Nevertheless, while these students had opportunities to talk together without teacher control, the teacher gave students tasks controlling, to some extent, the content of their talk. Moreover, all activities were followed up by a whole-class discussion where the teacher emphasised the points to be remembered.

This fifth criterion can be seen as a prerequisite for the other four of Englund's criteria – it has to be possible for students to deliberate outside teacher's "control". Practical work (in groups) in science works well in this respect as the students are working together with something concrete to talk about and teacher cannot be everywhere at the same time. However, a traditional conception of the established subject matter is perhaps narrowing the scope of the conversations for both students and teacher. From a research position, it should be possible, and interesting, to explore the outcome of teacher's facilitation of deliberative communication and identify students' further practices when guidance is withdrawn.

10.6 Discussion

In this chapter, we explore the possibility of researching citizenship, within ordinary science education, although such an approach is not present in current science education research. We have tentatively operationalised Englund's categories of deliberative communication and shown that it is possible to identify the presence and absence of relevant practices empirically. Thus, deliberative communication as an approach to citizenship education seems to be researchable also within ordinary school science. Moreover, this means that some practices reflecting citizenship education are present in ordinary school science, even if not stated explicitly and even if the quality of relevant practices of course might vary. Furthermore, we see it as important to stimulate citizenship as practice in school science and to apply a broad or thick notion of citizenship (McLaughlin 1992). Deliberative communication can be seen as an appropriate analytical tool for exploring the overlap between citizenship and students' practical work in school science. In that case, one might research the details of such practices and how they might be strengthened. Importantly, practices established in school might influence how students approach democratic issues (with a science component) later (Davies 2004).

Through the analyses, we found instances where students challenged views and made space for, listened to and respected each other's views and arguments. We also saw elements of collective will-formation and cautious questioning of teacher utterances. Several of these characteristics were present in the absence of the teacher. Furthermore, we exemplified instances where these characteristics were not fulfilled, for instance, lacking questioning of authority where this would have been appropriate as well as lacking respect for peers' contributions through "passive disrespect" and "mutual ostracising". We believe these findings supporting our claim that instances of deliberative communication are present in science teaching and possible to identify and make researchable.

In the beginning of the chapter, we argued that deliberative communication is coupled with a way of enacting citizenship, making an exploration of the overlap between citizenship and students' practical work in school science relevant. In addition, deliberative communication has many of the same traits as scientific knowledge building, coupling deliberative communication with science as a school subject. Scientific knowledge building has a distinct communicative side. One might say that knowledge that is not communicated does not count as knowledge (Ziman 2000). Moreover, it has been claimed that students need to work with and make explicit the "workings of science" (Lederman and Lederman 2014) to enable the use of scientific knowledge for civic purposes (Davies 2004). The arguments for this claim are that this is as vital as the facts and laws of science but also that an understanding of how scientific knowledge is built helps students to be (constructively) critical to science and its impact on society (Kolstø 2001). Deliberate communication shares many virtues with scientific communities. On an overarching level, one might say that they seem to share the valuing of argumentation and rational discussion. This is not to claim that each single participant in deliberative or scientific communication adheres to a particular set of virtues. However, it is possible to argue that within scientific communities, there exist virtues and standards which are appealed to in collective critical discussions (Longino 1990). Also, virtues related to deliberative communication might function as shared standards possible to appeal to, e.g. in situations of violation of these (Mansbridge et al. 2006). One common assumption is that arguments and discussions are important as they will ensure that empirical material and diverse points of view are assessed and that good argumentation prevails and leads to sound conclusions (Ziman 2000). In other words, the better argument "wins", at least in the long run. This model of reasoning can of course be critiqued – as it has been by several researchers (Kock 2007). For instance, an approach to argumentation that does not take into account power relations between the participants can be seen as simplified. Similar power-relation processes can be identified in the science classroom, as we also claim to have seen in this limited material: The students' impact on the joint work varies. Another shared virtue is the role of peers in will-formation. In professional science, peers play a vital role in establishing knowledge. Several avenues exist for the presentation and critique of views and arguments, such as journals, conferences and peer review processes. To commit the scientific work to serious scrutiny is essential in order to ensure quality. This can be seen as a form of collective will-formation and has historically resulted in canonical scientific knowledge. However, also here, there are critiques as claims have been made that gatekeepers can use their position to make personal or institutional interest influence the judgement (Ziman 2000). A last virtue we will explicate is the role of critique and questioning. The natural sciences have a proud tradition of questioning authorities within their own communities (Bjørkum 2009). According to Helen Longino (1990), although virtues and standards appealed to might differ between scientific communities, they will only be objective to the degree that they satisfy certain criteria for critical discourse. Amongst these are avenues for equality of intellectual authority. This implies that, in principle, it is not important who you are and where you come from – but rather, the strength of empirical and theoretical argumentation. This is of course, in reality, more complicated in science than within the framework of deliberative communication. However, as with deliberative practices, this norm influences practices and sanctioning mechanisms and might be seen as a counterpower to power relations threatening possibilities to question authorities.

Our conclusion is that virtues which are shared and possible to appeal to in scientific communities have several similarities with Englund's criteria for deliberative communication. The virtues involved in deliberative communication are therefore, in our view, legitimate to promote and practise in school science in general and in school science for democratic participation in particular (Kolstø 2005). Moreover, research on discourse practices in school science might also be compatible with deliberative communication, and it might be fruitful for the implementation of deliberative communication to look to research on discourse practices. Lastly, as we see it, quality indicators of deliberative communication provide the possibility to explore educational outcomes focusing on practices relevant to citizenship *and* the use of established subject matter in argumentation. According to Biesta (2009), the focus on "measuring" quality of educational outcomes that directs educational policy also seeps into educational practice. The system's quality measures, he claims, are to a very little extent dealing with the broader mandate of school – such as participation in democracy. The system for individual assessment of students might act contrary to argumentation and collective will-formation. Research on deliberative communication practices, and lack thereof, in ordinary science teaching and learning might provide one way to document the quality of practices and student competencies not easily measured by standard surveys. Moreover, research results on deliberative communication practices in ordinary school science might increase the focus on students as citizens in the school debates and in science teaching.

References

Afdal, G. (2010). The maze of tolerance. In K. Engebretson, M. de Souza, G. Durka, & L. Gearon (Eds.), *International handbook of inter-religious education* (pp. 597–616). Dordrecht: Springer.

Albe, V. (2015). Science for citizenship. In R. Gunstone (Ed.), *Encyclopedia of science education* (pp. 904–905). Dordrecht: Springer.

Bergqvist, K., & Säljö, R. (1994). Conceptually blindfolded in the optics laboratory. Dilemmas of inductive learning. *European Journal of Psychology of Education, 9*(2), 149–158.

Biesta, G. (2009). Good education in an age of measurement: On the need to reconnect with the question of purpose in education. *Educational Assessment, Evaluation and Accountability (formerly: Journal of Personnel Evaluation in Education), 21*(1), 33–46.

Birmingham, D., & Calabrese Barton, A. (2014). Putting on a green carnival: Youth taking educated action on socioscientific issues. *Journal of Research in Science Teaching, 51*(3), 286–314.

Bjørkum, P. A. (2009). *Annerledestenkerne: kreativitet i vitenskapens historie [Thinkers out of the ordinary. Creativity in the history of science]*. Oslo: Universitetsforlaget.

Crick, B. (2002). *Democracy: A very short introduction*. Oxford: Oxford Paperbacks.

Davies, I. (2004). Science and citizenship education. *International Journal of Science Education, 26*(14), 1751–1763.

Dewey, J. (2007/1916). *Democracy and education*. Teddington: Echo Library.

Dewey, J. (2015/1909). Moral principles in education. In *John Dewey. Combo Vol I* (pp. 223–253). Colorado: Springs Create Space Independent Publishing Platform.
Engle, R. A. (2011). The productive disciplinary engagement framework: Origins, key concepts and developments. In D. Y. Dai (Ed.), *Design research on learning and thinking in educational settings: Enhancing intellectual growth and functioning* (pp. 161–200). Oxon: Routledge.
Engle, R. A., & Conant, F. R. (2002). Guiding principles for fostering productive disciplinary engagement: Explaining an emergent argument in a community of learners classroom. *Cognition and Instruction, 20*(4), 399–483.
Englund, T. (2006). Deliberative communication: A pragmatist proposal. *Journal of Curriculum Studies, 38*(5), 503–520.
Englund, T. (2015). Towards a deliberative curriculum? *Nordic Journal of Studies in Educational Policy, 1*(1).
Flyvbjerg, B. (2001). *Making social science matter: Why social inquiry fails and how it can succeed again*. Cambridge: Cambridge University Press.
Fraser, B. J., Tobin, K., & McRobbie, C. J. (Eds.). (2011). *Second international handbook of science education*. Dordrecht: Springer.
Goldman, R. (2007). Video representations and the perspectivity framework: Epistemology, ethnography, evaluation, and ethics. In R. Goldman, R. Pea, B. Barron, & S. J. Derry (Eds.), *Video research in the learning sciences* (pp. 3–38). Oxon: Routledge.
Hamza, K. M., & Wickman, P. O. (2013). Supporting students' progression in science: Continuity between the particular, the contingent, and the general. *Science Education, 97*(1), 113–138.
Kelly, G. J. (2014). Discourse practices in science teaching and learning. In N. G. Lederman & S. K. Abell (Eds.), *Handbook of research on science education*. Oxon: Routledge.
Knain, E. (2005). Identity and genre literacy in high-school students' experimental reports. *International Journal of Science Education, 27*(5), 607–624.
Kock, C. (2007). Norms of legitimate dissensus. *Informal Logic, 27*(2), 179–196.
Kolstø, S. D. (2000). Consensus projects: Teaching science for citizenship. *International Journal of Science Education, 22*(6), 645–664.
Kolstø, S. D. (2001). Scientific literacy for citizenship: Tools for dealing with the science dimension of controversial socioscientific issues. *Science Education, 85*(3), 291–309.
Kolstø, S. D. (2005). Hvilke verdier er dannende i naturfaget? [Which values are educational in school science]. In K. Børhaug, A.-B. Fenner, & L. Aase (Eds.), *Fagenes begrunnelser. Skolens fag og arbeidsmåter i danningsperspektiv [Arguments for school subjects. School subjects and ways of working in a Bildung perspective]* (pp. 47–66). Bergen: Fagbokforlaget.
Lederman, N. G., & Lederman, J. S. (2014). Research on teaching and learning of nature of science. In N. G. Lederman & S. K. Abell (Eds.), *Handbook of research on science education* (pp. 600–620). New York: Routledge.
Levinson, R. (2010). Science education and democratic participation: An uneasy congruence? *Studies in Science Education, 46*(1), 69–119.
Longino, H. E. (1990). *Science as social knowledge: Values and objectivity in scientific inquiry*. Princeton: Princeton University Press.
Mansbridge, J., Hartz-Karp, J., Amengual, M., & Gastil, J. (2006). Norms of deliberation: An inductive study. *Journal of Public Deliberation, 2*(1).
McLaughlin, T. H. (1992). Citizenship, diversity and education: A philosophical perspective. *Journal of Moral Education, 21*(3), 235–250.
Rienstra, B., & Hook, D. (2006). Weakening Habermas: The undoing of communicative rationality. *Politikon: South African journal of political studies, 33*(3), 313–339.
Roberts, D. A., & Bybee, R. W. (2014). Scientific literacy, science literacy, and science education. In N. G. Lederman & S. K. Abell (Eds.), *Handbook of research on science education* (pp. 545–558). Oxon: Routledge.
Thomas, G. (2011). *How to do your case study. A guide for students and researchers*. London: Sage.
Utdanningsdirektoratet [Norwegian Directorate for Education and Training] (2013). *Læreplan i naturfag [Curriculum for School Science]*. Retrived from http://www.udir.no/kl06/NAT1-03/
Ziman, J. (2000). *Real science: What it is and what it means*. Port Chester: Cambridge University Press.

Chapter 11
In-Between Chapter: Troubling the Social – Entanglement, Agency, and the Body in Science Education

Marianne Løken and Margareta Serder

11.1 What Is the Social that We Trouble?

This introduction to the "social" in science education that is troubled in the following texts will be based on our own experiences of adopting various socio-material theories in research. These experiences have brought us to recognize the challenges that emerge and which we wish to address in this in-between chapter.

First, what is *the social* that we trouble? To start off, we might choose to translate "social" into "human", thus understanding it as involving all endeavors where humans are. To us, however, troubling the social in science education means to accommodate its interconnectedness with all "the rest". Consequently, the social needs to be understood not as an entity of "human" but in combination with its entire social, cultural, and material embeddedness. We call this a socio-material standpoint (Serder 2015; Løken and Sørensen in review), which, in this sense, leans toward post-humanist material feminist theorizations and learns from science and technology studies (STS) and actor-network theory (Latour 1999; Law 1994). The purpose is to assert the principle that entities take shape through interaction or, as perhaps more precisely termed by Karen Barad (1998), "intra-action" with other entities. In our view, one of the most important benefits of this research approach is the troubling of the question of agency in general and, more specifically, "the possibility of non-human forms of agency" (Barad 1998, p. 112).

To be human is to have ongoing relations with everything nonhuman or, to paraphrase Bruno Latour in his *Reassembling the Social* (2005, p. 2), "It is no longer

M. Løken
Norwegian Directorate for Education and Training, Oslo, Norway
e-mail: marianne.loken@udir.no

M. Serder (✉)
Skåne Association of Local Authorities, Malmö University, Malmö, Sweden
e-mail: margareta.serder@kfsk.se

K. Otrel-Cass et al. (eds.), *Cultural, Social, and Political Perspectives in Science Education*, Cultural Studies of Science Education 15,
DOI 10.1007/978-3-319-61191-4_11

clear whether there are relations that are specific enough to be called 'social' and that could be grouped together in making up a special domain that could function as a society." In the following three texts, we claim that the relations between humans and nonhumans, and indeed between humans themselves, are the core focus of interest. Hopefully, the reader will find that such an understanding of "the social" allows the researcher to raise questions about learning and knowing.

11.2 A Socio-material Approach

We have employed a socio-material approach in this in-between chapter by engaging with the following three texts as empirical research material with which we are intertwined. This has entailed a method of going back and forth in dialogue with each other and the texts themselves—the materials with which we entangle. This has given us an insight into the method of *diffractive reading* (Barad 2007), which is used in one of the texts by Auli Arvola Orlander and Marie Ståhl, by actually performing the method ourselves during 3 days together. Just as with Arvola Orlander and Ståhl, we believe to have experienced *becoming with the data* (Hultman and Lenz Taguchi 2010).

When we started to engage in socio-material theories, one of the most difficult ideas was how can we consider nonhuman matter in terms of agency. In two of the texts, the nonhuman takes the form of a human with nonhuman extensions, by using Donna Haraway's figuration of the *cyborg* (Haraway 1991). The cyborg, as with other figurations, allows us to think differently about how things are put together (Haraway 1991). We conceive of figurations as a way to approach the possible agency of artefacts, molecules, technologies, and even rituals and practices. Furthermore, we understand agency as an effect of intra-actions between the human and nonhuman. According to Barad, agency is not something humans or nonhumans have but what occurs in the actual engagement or intra-action (Barad 2003). Therefore, another word for intra-action is co-constitution.

A socio-material approach seeks to avoid representational views of knowledge and replace them with other ways of thinking about objectivity and truth (Serder 2015). This will be evident to the readers of the three texts that constitute our empirical material. The reader will also become aware of the three authors' struggle to be fair to the social and material aspects in the analysis of their data. We have also experienced this ourselves. The question is: How can these appealing theories operate with the empirical material in order to serve them well? We are using this opportunity to stress what is in it for us, that is, to articulate some of the contributions of the socio-material stance through examples of how the authors have engaged with this task.

11.3 The Material Body in Science Education and Research

The three texts do not unanimously adopt an explicit socio-material approach; however, two have an explicit post-human stance, namely, the chapter by Arvola Orlander and Ståhl and the chapter by Ståhl. Kathrin Otrel-Cass actually defines her research as sociocultural with references to John Dewey and Lev Vygotsky. However, in all of the texts, we can read an interest in becoming with the embodied material world. Otrel-Cass addresses the sensory aspects of science education learning and the interplay between the senses and the material world. The inseparability of the mind from the body is likewise recalled in her text, as she refers to Dewey's term "body/mind" in order to bring the two concepts closer together. The body/mind is present in all of the texts and makes room for new relations and interpretations in science education research. Otrel-Cass writes about "seeing" with her ears, and Arvola Orlander and Ståhl are reading "with" their bodies. The latter two authors claim that they have experienced learning in a bodily effective way. Knowing thus emerges "in between", something that occurs for the authors during their diffractive reading inspired by approaches often labeled as "material feminism" (see, for instance, Alaimo and Heckman 2008).

According to Stacy Alaimo and Susan Heckman, "material feminists want to know how we can define *the real* in science and how we can describe non-human agency in a scientific context" (2008, p. 7). In Andrew Pickering's work (1995), the human and the nonhuman in scientific practice are in constant and mutual accommodation and resistance, in a "dance of agency" (1995, p. 102), a movement for which he uses *the mangle* as a metaphor. The mangle pulls material agency into the terrain of human agency and mixes everything up, such as Arvola Orlander and Ståhl interestingly try to do in their work on diffractive reading.

Using the concept of the "mangle" to explore practice and science education issues can be particularly useful, as it allows the researcher to understand the array of causal factors involved in the event (see, for instance, Serder and Ideland 2016). Ståhl uses "mangling" as an analytical tool in her reflection of how the intra-actions between a chemistry test and the students taking it produce different subject positions. With inspiration from Haraway's concept of "the apparatus of bodily production" (Haraway 1991), Ståhl wants to emulate the test as an apparatus—a world in the making (Haraway 2004, p. 330)—allowing for different figurations to emerge from the mangling. It appears to us that, also in Otrel-Cass' text, we could use the concept of apparatus for the sensory pedagogy she is advocating, using such examples from students' experiences as hot/warm water, glasses, and plastic wine corks. Nevertheless, the material is more than a tool or an artefact. The apparatus as a feminist figuration is a relation: an intra-action.

Feminist figurations, used analytically in the texts of two of the authors, speak to thought, feelings, and body (Lykke 2009). Why involve the material body as sensoriality and affections in research? The answer could simply be: Why not? Feminists have critiqued the scientific rationalist discourse and the nature/culture dichotomy for decades. However, in an antiessentialist mode, the objectification of bodies has continued to trouble feminist thought. Haraway's concept of the "God trick" relates to the problematic idea that nature and bodies are passive resources for culture and mind (Haraway 1991), to which Arvola Orlander and Ståhl respond: "The body has to be involved because the body is always a complex, contradictory, structuring and structured body whose parts are impossible to separate" (Arvola Orlander and Ståhl, referring to Haraway 1988, p. 589). The authors continue by concluding that the only position from which objectivity cannot be practiced is from the "standpoint of the master", illustrated by Haraway's metaphor of the God trick.

11.4 Disclosure of the Social from a Post-humanist Stance

The commonality of these three texts is the troubling of reductionist analysis or rather some of the ideas that are often attributed to a positivistic scientific stance: the goal of science as rational, logical, disembodied, objective, and value-free knowledge. They also have in common a more or less explicit critique of dichotomies, whether it is the body/mind, nature/culture, or human/nonhuman. Our experience is that although postmodernism as a whole has given us important insights, it is also the case that some postmodern thinkers do the opposite of what postmodernism claims to accomplish, that is, to deconstruct dichotomies. The challenge that confronts postmodern and post-humanist thinkers, including ourselves and the authors of the three chapters, is to define a theoretical position that does not privilege either the human or nonhuman but instead explores their intimate co-constructions (Løken and Sørensen in review).

In this book, the cultural, the social, and the political have been separated into three different perspectives. Our "socio-material reflex" suggests integration to dissolve categories such as these. How otherwise can we deal with the hybrids and tricksters (Haraway 1991)? At this point, we have arrived at one of the ever-present risks of reproducing the same dichotomies and pedagogies of difference that the material feminist or post-humanist-inspired researcher is actually seeking to escape: While problematizing *some* categories, their opposites are simultaneously construed. We believe it also illustrates how constrained we are as researchers operating within our human discursive limits.

This in-between chapter has examined how a socio-material approach refuses to take for granted the distinction between the social and cultural, and the human and nonhuman, and moves beyond the zones of human-centered research at large. Our objective has been to contribute to a new understanding of how reading the three texts with a socio-material approach draws the material back in. By learning from the work of Haraway, Barad, and other post-humanists about the complexity of our world and becoming with, we can work from here to develop reflections that speak across differing forms of becoming with in science education research.

References

Alaimo, S., & Heckman, S. (2008). *Material feminism*. Bloomington/Indianapolis: Indiana University Press.

Barad, K. (1998). Getting real: Technoscientific practices and the materialization of reality. *Differences, 10*(2), 87–128.

Barad, K. (2003). Posthumanist performativity: Toward an understanding of how matter comes to matter. *Signs: Journal of Women in Culture and Society, 28*(3), 801–831.

Barad, K. (2007). *Meeting the universe halfway. Quantum physics and the entanglement of matter and meaning*. Durham/London: Duke University Press.

Haraway, D. (1988). Situated knowledges: The science question in feminism and the privilege of partial perspective. *Feminist Studies, 14*, 575–599.

Haraway, D. (1991). *Siamans, cyborgs and women. The reinvention of nature*. New York: Routledge.

Haraway, D. (2004). *The haraway reader*. New York: Routledge.

Hultman, K., & Lenz Taguchi, H. (2010). Challenging anthropocentric analysis of visual data: A relational materialist methodological approach to educational research. *International Journal of Qualitative Studies in Education, 23*(5), 525–542.

Latour, B. (1999). *Pandora's hope: Essays on the reality of science studies*. Cambridge, MA: Harvard University Press.

Latour, B. (2005). *Reassembling the social: An introduction to actor-network theory*. Oxford: Oxford University Press.

Law, J. (1994). *Organizing modernity*. Oxford: Blackwell.

Løken, M., & Sørensen, S. Ø. (in review). *Materialiteten "kicks back": Om betydningen av materielle praksiser og erfaringer i forståelsen av kjønnede utdanningsvalg* [Tentative title].

Lykke, N. (2009). *Genusforskning: En guide till feministisk teori, metodologi och skrift*. Stockholm: Liber.

Pickering, A. (1995). *The mangle of practice: Time, agency and science*. Chicago: University of Chicago Press.

Serder, M. (2015). *Möten med PISA. Kunskapsmätning som samspel mellan elever och provuppgifter i och om naturvetenskap*. [Encounters with PISA. Knowledge assessment as co-action between students and scientific literacy test items]. Malmö: Malmoe University. Dissertation.

Serder, M., & Ideland, M. (2016). PISA truth effects: The construction of low performance. *Discourse: Studies in the Cultural Politics of Education, 37*(3), 341–357.

Chapter 12
Towards an Understanding of Diffractive Readings of Narratives in the Field of Science Education

Auli Arvola Orlander and Marie Ståhl

12.1 Introduction

We have been involved on several occasions where the research, rooted in science, calls for an objective researcher position. A mission of that it is possible to observe; a researcher sitting in a corner observing what is going on "out there" without direct involvement; a white Western researcher with colonial connotations objectively describing the true happenings. This is a view based on binary thinking and a dichotomous framework. For instance, this objectivity excludes embodied feelings. Instead, in the study we wanted to challenge this ontological view since it disregards humans to be both mind and body, and that research is always situated in a context. At the same time it made us think: Can we place ourselves outside the box of what has been learnt as "proper" research behaviour and not hold back? To let it go and to show emotion, to become part of the world, engage with the data, to become affected by it, to become part of it and to actually be in the world and to feel the reaction from the whole of one's body when conducting research.

During a period of time, in search of different kinds of encounters with the data, we have drawn on the writings of Donna Haraway and Karen Barad. We have returned to them on various occasions with the aim to understand teaching/education in relation to theories of diffraction. To be frank, it has not been easy. It has been a struggle to understand and to position ourselves differently. At the same time the idea about the objective, rational and logical scientist without feelings has always been, in our opinion, an impossible mission. Thus, Haraway and Barad have

A.A. Orlander (✉)
Stockholm University, Stockholm, Sweden
e-mail: auli.arvola.orlander@mnd.su.se

M. Ståhl
University of Gothenburg, Gothenburg, Sweden
e-mail: marie.stahl@gu.se

K. Otrel-Cass et al. (eds.), *Cultural, Social, and Political Perspectives in Science Education*, Cultural Studies of Science Education 15,
DOI 10.1007/978-3-319-61191-4_12

helped us think along new lines and have been an important inspiration for many feminist researchers (see, for instance, Taylor and Ivinson 2013). It is about theories that attract and affect. Encouraged by Haraway's and Barad's writings, we decided to embrace their view of objectivity and their analytical tool, diffractive reading, to explore the outcomes when analysing students' texts in chemistry as well as an excerpt from an observation in science class.

After an introduction to our understanding of key concepts in post-human approaches, we will discuss examples from our own research. This will show how material feminist theories can influence our discussions and outcomes when reading data in search of different kinds of encounters. Accordingly, this study is a process-oriented study.

12.2 Challenging Objectivity and Cartesian Thought

In this study we try to identify other ways of exploring empirical data than the traditional positivistic approach. We challenge both the positivist approach to research and its view on objectivity, as well as the natural science embrace of the Cartesian tradition of thought. The latter portrays science as being objective, logical and rational as opposed to subjective, emotional and irrational. It also embraces the dichotomous notions regarding body-mind, woman-man and nature-culture relations (Haraway 1991; Brickhouse 2001).

The value-neutral norms for scientific research, as reproduced in scientific education (Östman 1995; Svennbeck 2004), means that the researcher should be objective and take an independent position towards the examined object. This also applies to students when those in the chosen examples discuss scientific issues. The objective aim is also reflected in how scientific texts are written. By using passive tense and turning verbs into nouns, the text becomes tinglified, and the real agent (implicitly understood as the human) hidden (Halliday 1996). According to Barad (2007), it is not only the human agent that becomes hidden, but also the nature, the body and all other matter become dead and irrelevant. This also means that objectivity is placed in the head and in a bodiless intellect that lacks the expression of emotions. However, emotions are necessary. The body has to be involved because the body is always "… a complex, contradictory, structuring and structured body …" (Haraway 1988, p. 589) whose parts are impossible to separate. This means that research and learning is always situated in a context that includes the agent's own body (Haraway 1991; Milne and Rubin 2011).

Similarly, Barad (2003, 2007) argues that the researcher is part of the scientific community and therefore is unable to stand free from her or his empirical data and apparatus. Accepting Barad's approach entails that ethics and policy issues must always be part of scientific work.

Intense scrutiny, criticism and questioning of the research subject and discovered context are the basis for credible objectivity (Haraway 1989, 1991). "Identity, including self-identity does not produce science, critical position does, that is objectivity" (Haraway 1991, p. 193). The only position from which objectivity cannot be practised is from "the standpoint of the master" (Haraway 1991, p. 193), the so-called God trick.

There is not a single feminist position, but the goal is the same: to get a better scientific description of the world and that "The science question in feminism is about objectivity as positioned rationality ... of views from somewhere" (Haraway 1991, p. 196). With the God trick, as the positivist objectivity is interpreted from a feminist perspective, the subject (the researcher) and the object (research object) are related to each other in a "rule-slave" relationship. This is manifested in research situations by the anthropocentric relationship between man and nature and other dichotomous conditions, as the Cartesian tradition of thought represents. The man and the natural sciences, and their common objective features, are the norm and stand above nature and non-male humans.

12.3 Theoretical Tools

In search of a view, we have used Haraway's and Barad's thinking tools, diffractive reading and the cyborg theory to explore the data. This has meant stepping down from our privileged human position, to dismiss the gaze from above (Haraway 1991) and instead emerge to an event within the data, i.e. a diffractive materialistic reading. Diffractive reading is to experience an instantaneous encounter and entanglement with the data. Not before or after the encounter, but in the moment. This also means that in order to understand, we (the researchers) and the data are no longer objects with inherent boundaries and fixed properties. Instead, bodies are always attached to specific meanings in specific events of intra-activities.

12.3.1 Diffractive Materialistic Reading

Diffractive materialistic reading is a thinking tool for the critical analysis of, for example, a text (Barad 2007; Haraway 1997). This concept emanates from Haraway's and Barad's further development of the concept's metaphorical meaning to identify an object of research by diffraction instead of reflection. The latter is otherwise the usual way to describe the processes of interpretation. In physics, diffraction is the phenomenon that occurs, for example, when studying light. Unlike what occurs at a reflection when the light from the object is reflected in a mirror and then thrown back towards the viewer, diffraction is obtained when light passes through a grid, which then distributes the light in different ways depending on the grid design. Various large gaps create a different spread of the light passing through and cause the different colours or patterns to form on a screen/wall that captures the diffraction. Haraway (1997) explains that seeing the analysis as a reflection maintains the object being studied, while through diffraction a change is made that not only includes the item itself but also its foreground and background. This materialist discursive practice can be understood with the help of Haraway's string figures (Haraway 1994, p. 69): a game many of us played when we were young. It entails that a string is tied together and wrapped around one participant's thumb and little finger on each hand, forming a pattern. New patterns are created when the string changes hands to a new participant. In Haraway's interpretation it becomes a metaphor for how patterns develop and change while switching actors and thus how the background and foreground are constantly shifting (Haraway 1997). The story performed can therefore be replaced by another possible story, and the constructed storyline can be analysed in different ways to mean something completely different (Jackson and Mazzei 2012). This approach sees various types of displacements of the object's foreground and background, which may then see different aspects of the imploding dimensions come forward (Lykke 2009).

Diffraction can metaphorically be described as waves breaking against a solid object (Barad 2007). The diffraction occurs when the wave breaks up into smaller waves and diffuses out of each other. Agency occurs in this encounter, "*in-between* [sic] different bodies involved in mutual engagements" (Hultman and Lenz Taguchi 2010, p. 530). This entanglement of intra-action agencies is understood as a phenomenon or as "the ontological inseparability/entanglement of intra-action "agencies"" (Barad 2007, p. 139). Agency is not something that humans or non-humans have but occurs in the actual engagement, i.e. the diffractive entanglement. Yet it only occurs at that moment (not before or after) and only together with other components. Barad calls this phenomenon an intra-action: when the joint formations of entrapped agents are formed. This is unlike an interaction where each component is presumed to exist as an individual element both before, during and after the encounter (Barad 2007).

When we use diffractive reading we are looking for events, activities and meetings that cause changes in the components of an intra-activity, a "... diffractive way of seeing ..." (Hultman and Lenz Taguchi 2010, p. 535). One cannot, as is usual in research, take a step back and reflect on what one sees and thus be outside the phenomenon. On the contrary, it is about installing "… yourself into an event of becoming-with the data" (Haraway 2008, p. 16). It means using all of our affective and cognitive abilities. However, it is not a question of trying to assume the role of student or co-author in an attempt to understand what was going on at the time the text was written, "while engaging in diffractive 'seeing' and 'reading' with it" (Hultman and Lenz Taguchi 2010, p. 537). It is about making a description of the text on the basis of the meeting that takes place between the researcher and the text. In this meeting the researcher must shield his or herself from everything that is otherwise taken for granted. Karin Hultman and Hillevi Lenz Taguchi (2010) argue that diffractive reading can also be understood as nomadic thinking (Deleuze 1990). In turn, this means that we have to put off the yoke that makes us think in dichotomous units and instead opens up for another encounter of knowledge that can be interpreted as a sign, resulting in bodily affective learning. Here the knowing emerges in-between producing another kind of knowing (Hultman and Lenz Taguchi 2010). This means that we go beyond the human/non-human divide and acknowledge our coexistence with the rest of the world. Parallels can be drawn to Barad's thesis that we cannot separate knowledge from being "...we know because we are of the world" (Barad 2007, p. 185).

Diffractive reading is an onto-epistemological study, which means that it is a "... study of practices of knowing into being ..." (Barad 2007, p. 185). Hultman and Lenz Taguchi (2010) use the word enchanted, but we would rather use the Swedish word *förundrad,* which is close to *marvel at* in English. It describes dimensions of the feeling and commitment that exists when we are deeply part of an intra-activity in the world.

12.3.2 Cyborg Theory

We also draw on cyborg theory developed by Haraway (1991) as a positive feminist figuration that blurs the boundaries between the human and the non-human: a fusion of body and technology to liberate us from the stereotypical images that we are divided into. Haraway claims that we are all cyborgs and that the boundary between fiction and reality is an optical illusion (Haraway 1991). The concept of the cyborg also includes a trickster. A trickster exists in all matter and cannot be controlled, appearing and making fun of us, and works against the anthropocentric gaze from above that neglects most of the world since only the human world counts as real and active. (An effect that happens in our bodies as well as in other matter.) The trickster is what Haraway calls feminist objectivity. Something that gives the opportunity for surprises, since we cannot rule the world and control what will happen. We are simply living here (Haraway 1991). Although researchers create knowledge about the world, one cannot ignore the fact that the universe also kicks back (Haraway 1991). Therefore, matter has agency in its intra-acting and entanglement with other matter. Trying to use the *God trick* is useless, however, since "We do not obtain knowledge by standing outside the world; we know because 'we' are *of* the world" (Barad 2003, p. 829).

12.4 Working with Agential Cuts

The two studies on which we draw are situated in a Swedish educational context and consist of grade-9 students' answers collected from Swedish national tests in chemistry, as well as an excerpt from classroom observations in upper secondary schools. These may be regarded, through Barad's conceptualization, as two agential cuts from reality (Barad 2007, p. 148).

Barad means that "Apparatuses enact agential cuts that produce determinate boundaries and properties of "entities" within phenomena, where "phenomena" are the ontological inseparability of agentially intra-acting components" (Barad 2007, p. 148). This means that a localised subject position where the researcher – as well as the object to be examined, languages, devices and everything else – is inextricably part of the cut of the reality that is made. In other words, the researcher does not stand outside the object, approaching it with a representational view, but is part of it and entangled within (Barad 2007). Therefore, the researcher is always in a special material-discursive practice. Barad means that discursive practices are material prerequisites for meaning or limit-setting practices that have no end in an ongoing intra-activity; thus, "discourse is not what is said; it is that which constraints and enables what can be said" (Barad 2007, p. 146). Furthermore, an agential cut construes objective knowledge. If one fails to make cuts and define the boundaries, then it is impossible to work scientifically, Barad (2003) claims. Unlike traditional epistemology, which works with a universal cut between the world and object, the researcher constructs the object.

Aim and Research Questions
Our aim is to discuss the findings in the post-human approaches applied to our agential cuts. The question we are investigating is:

- How is knowing produced differently in our diffractive readings?

  With some follow-up questions:

- The researcher as a performative agent doing agential cuts: What kind of researcher subjectivity does this entail?
- How/when is this enactment of diffractive analysis also an enactment of ethics?
- In what way can diffractive reading enact/contribute to our understanding of the excerpts?

12.4.1 About Us

The participants in the construction of the apparatus of knowing are two white, middle-class, female, science education researchers who live in Sweden. We have both been active science teachers in Swedish primary schools. We also draw our empirical examples from this subject area. For our common diffractive reading, each of us selected a short section of the text taken from a major empirical data material. In the shared process, we had the desire to achieve critical and renewing discussions of our respective empirical data, in which we review agential cuts (Barad 2007). The first cut concerns a student's answer to an item concerning discoveries in chemistry and its impact on society. The second cut is a classroom situation where students debate for and against nuclear power as an energy source. Both examples had been previously analysed and resulted in an analysis of teacher and student actions, while downplaying the researchers' roles. The presentations of the results were tangled in a dualistic view, in a traditional way of presentation, where the educational situations were more or less disconnected from us as researchers.

12.5 Procedures

We have encouraged the discussion flow from everything between earth and heaven and all that came to mind when we read the texts. We hooked on and surfed each other's stories, letting the story adopt the form it did in that moment and allowing it to float back and forth. In a later phase, we discussed more comprehensively about what we had been through and what new images had produced in our diffractive readings. We also discussed the differences to previous readings.

The focus here is to describe the scientific process we became involved in when using diffractive reading. Accordingly, we will not treat the specific research questions derived from our two earlier separate studies (mentioned above).

12.6 Results of the Diffractive Reading: What Connections, Interferences and Sensations Emerge, or are Evoked, in Our Readings

In this section the content of our two diffractive readings, one at a time, are presented in more detail. First, the context of the studied examples is presented and is followed by an account of the diffractive reading.

Example 1 Discoveries in Chemistry

The following text is taken from the Swedish national test in chemistry carried out each year in grade 9. The text is a student's response to one of the items included in the 2009 test. The specific item is formulated in the following way: Choose a discovery in chemistry and describe how this discovery has changed humans' ways of thinking and doing things.

Student's response:

> Something very topical that has been current a long time is health. Humans have ... [not possible to decipher] ... things to feel good. The discovery of vitamins came about that researchers noticed that some deficiency diseases could be cured with certain foods. Obviously, no one said "aha" the calcium in milk reduces the risk of osteoporosis, but the notions of vitamins began. The discovery of how carbohydrates, fats and proteins in the body function has changed millions of people's way of life. The fact that new diets is constantly brought up by burning fat, by eating fat, building muscles, by eating proteins and increasing endurance by eating slow carbohydrates then the hysteria to get the dream body continues even if some diets are pure madness.
>
> The diet of Atkins, e.g. aims to exclude carbohydrates or almost only eat proteins. Although Atkins himself died of elevated levels of blood fats, many comply with the Atkins diet. Articles that newspapers publish with complex concepts and complex ideas ... it is so elaborate and complex, so it must match according to some. The Atkins diet is just one of many methods, but the discovery of how carbohydrates, fats and proteins function in the body has led to such a frenzy that one now listens to what the articles say, instead of the body's own unique needs. The soul, it is now backed into a corner somewhere.

Diffractive reading 1: A student response to the Swedish national test in chemistry.

The escape from the concrete text was completed in small steps. The body, however, was the focus for both of us. We saw the "big fat" and the "big muscular" bodies. They were placed in a gym, where they appeared mechanically following the movements from the equipment. Static and repetitive movements developed: sweat dripping and glowing skin. Fat and protein was embodied in the partly grotesque, disgusting rolls of fat and as protein in muscle, which in turn became bloody steaks. They also gained colour, substance and eventually odour. Associations were also made to the inflated girl in the Chocolate Factory (film) that burst and whose stomach contents emptied out. Into these perceptions newspapers were mixed in, screaming about anabolic steroids and diets. The opposite appears as beautiful and graceful ballet dancers who push their floating bodies into bodily perfection and elegance and to "real-made men": the grandfather's muscular body made from hard work as a farmer. Also, beautifully thick bodies appear in Boleros paintings as well as ordinary thick, thin and calloused bodies in their plainness. The dichotomous bodies became prominent, the good, graceful ballet dancers and the farmer with the hard-working muscular body against the disgusting, fat or artificial body builders and the dichotomy between the muscular man's body and the obese female body. Our heterosexual beliefs became apparent.

However, the bodies lack souls. Sitting displaced and crouched outside the gym. The soul is genuine, it is not embodied, although she is free from moral disgust, a materiality that has been left outside our decadence. We moralise. The ballet dancer feels as free as the soul, but here the soul is sadly left on its own. The soul is seen as a woman and the discoverer of proteins, carbohydrates and calcium as a man.

Example 2: A Nuclear Power Debate

The second agential cut we processed had a slightly different character than the first one. It also proved to be more difficult to obtain a feeling of success in the question of opening up for new images.

This time the process concerns a nuclear power debate/argumentation in the classroom at an upper secondary school. Students have the task to play predetermined roles under the theme *for* or *against* nuclear power. The teacher begins by encouraging the class to compete.

Teacher:	The next thing we will do and that is examined, made sort of an assessment on, is a debate. Who is going to win it?
Students:	Me.
Teacher:	Who's going to win? [shouting now]
Class:	MEEEE!!!!

After half a day of preparation, with students searching for information online, the class have a debate. One group is sitting beside a table on the left side of the teacher with the role to advocate nuclear power. On the right side of the teacher sits the group with the role of arguing against nuclear power. The situation is described in the introductory part of the text we are working with, followed by a part where the teacher presents the grading criteria. This paper is held in the teacher's hand that students received earlier (these excerpts are not shown here) and at the end follows a short excerpt with students' arguments:

Jacob: It's sort of the best.
Nathalie: Around 50% of our energy comes from nuclear power.
Sitara: Secure technology for electricity production.
Geoff: Nuclear power is not a secure source of energy.
Jacob: Why?
Geoff: The waste is not so good.
Teacher: Develop what you say now... this not being secure.
Geoff: There is no guaranteed security for long-term storage of nuclear waste. If it becomes a... causes great problems for nature and the environment.
Jacob: Yes, but how do you think then that we will get all the energy, for 50% comes from nuclear power stations?
Geoff: Almost half of Sweden's energy is hydropower and we should rather invest in renewable energy sources.
Nathalie: Like what?
Geoff: Wind power.
Jacob: Hydro power?
Geoff: Hydro power, geothermal, others.
Jacob: What should we do in those countries where they do not work with, for example, hydropower?
Sitara: They have no clean water to drink.
Jacob: What will you do with uranium waste?
Teacher: Please go on ... develop a bit more. You, who are against. Why do you think the disposal of nuclear waste is not quite okay? Can you develop a bit more?
Jacob: It destroys ... uranium.
Teacher: How?
Geoff: When we store the waste, we cannot have secure storage. The waste can spread in nature. It is not safe.

Diffractive reading 2: A nuclear power debate.

Excerpts/data have been involved in another research process using feminist theories, but without Barad's theories regarding materialism. In other words, one of us has a previous relationship to the content. The other has faced the text for the first time; therefore, it was particularly interesting to hear the waves she surfed or hooked on. The opening phrase became "I feel stressed when I read this, hunted up by a mood of bustle/rush. The teacher wants the students to compete against each other, one side must win…".

We are reminded of the pressure of grading. The content feels unengaging and discussions meaningless. Students are throwing various allegations in a ping-pong style match but without playing with each other. Nobody listens to what the other party says, and it becomes callous robot-like statements. Nothing personal is included, and the whole situation feels meaningless. Our discussion appeared to idle, repeating ourselves in a way that reminds of prior proceedings.

Only when we affirmed our emotions the content began to move. We asked: What is it that makes us so angry? An outrage that was familiar from earlier processes with other researchers (those with a past as teachers). We were hit by illustration, reminding of the structures schools are stuck in and part of. In the debate, facts were thrown back and forth as pros and cons in question of nuclear power. Distanced, cold, masculine content. It was terse and uninspired, objective and rational. Emotional expressions were rare. The teacher explained the rules of the game: The more sources of facts, the better the chance of higher grades. Accordingly, the students searched for facts to be able to regurgitate them, in the name of assessment. The whole debate and the content of the discussion felt distant from a "real" experience and from students' everyday lives. We have both worked as teachers in schools; our "pedagogues" became upset by the feelings this example of "school reality" evoke. Our stories moved to the school premises where we once worked as teachers. Images of alternative teaching methods were created between us. The teaching should not be "this and that". Strong values evoke about how it should be managed and with "a touch of" idealised images of how we in our teaching deed would have differed in such a classroom situation.

It became clear that we engaged with the data in a trans-corporeal process. We were back in the classroom and a part of it. In the story the structure and curriculum content become part of materialisation. The classroom rating matrix the teacher held in his hand, facts and our memories/bodies/classrooms materialised in intra-action of the school structures.

12.7 Theoretical Associations

In this section we make an effort to theoretically describe what we have experienced when doing the diffractive readings, using the theoretical framework of diffractive reading together with examples from our entanglement with the excerpts.

According to Barad (2007), when completing diffractive reading, we and the excerpts are interfering with each other in an intra-action. In that very moment of entanglement, agency occurs among us, and the divide between the non-human and the human ceases to exist (Barad 2007). The phenomenon that the non-human has agency is a new experience to us but becomes apparent when words are no longer just words: they become material bodies that are interfering and making themselves intelligible to one another. At that time, we understood our bodies and the excerpts to be "… as a space of transit, a series of open-ended systems in interaction with the material-discursive 'environment', diffractive analyses [that] constitute transcorporeal engagements with data" (Lenz Taguchi 2012, s. 265).

The text is no longer a student text that we try to interpret but has received material dimensions in the intra-action, in-between our material bodies, when interfering and making us intelligible to one another. This can be viewed as metaphorical with help from Haraway's (1991) string figures; each one of us brings forward *different stories/new patterns* each time we hold the string, thus still entangled with the previous figures. For example, in the first diffractive reading, the focus on discoveries in chemistry have been replaced and are, in our entanglement, about how the bodily expressions affect our notions of being human. The ethical dimension is thus clarified in diffractive reading and has participated in the interpretation of the text in the sense of how our dichotomous, heterosexual perceptions, as well as our prejudices about the human body, have been highlighted and become apparent to us. The trickster has shown itself in these prejudices that we thought we did not have. Bodies doing static and repetitive movements that have developed sweat, dripping and glowing skin. The bodies in the text also became apparent in our release of the mind to the material creations of meat, fat and blood. In the gym, the human body – together with the machines that were used – became cyborgs, machine humans (Haraway 1991).

In the second diffractive reading, the rating matrix, scientific facts, classrooms, past-time memories as teachers and bodies materialise in intra-action of the school structures. A trans-corporeal process (Alaimo 2008) in which we engage with the data, into the textual discursive practice the excerpt consists of, became clear. We were back in the classroom and a part of it. The data "... interferes with the sensibilities of our body and minds what this brings to the event of reading the data" (Lenz Taguchi 2012, p. 272). The story, the structures and the curriculum content received material dimensions. It is clear that we did not only respond to the situation in itself but to the structures and discourses, which the whole school system is materialising: the political, the structural and the physical. In the second reading, the school discourse became an agent and limited what becomes possible to say, a discourse constraining and enabling what can be said (Barad 2007).

Haraway's cyborg theory (1991) illustrates how dependent humans/we are on technology, not only because it becomes clear in the first diffractive reading – to shape or reshape our bodies whether it is supposed to be natural or unnatural – but also how entangled we are in our reading to the understanding of what a body is and should be. The norm of what should be allowed concerning a shaped body, and what is not, stands out as something fundamental in our thinking as well as the divide between good and not good bodies.

The division of good and not good also occurred in the second reading, when our aim to denounce the pedagogy made us end up in the dichotomy of good/bad teaching. Dichotomies such as real/unreal were also brought to the fore when relating to the students' so-called real experiences in terms of the content of the debates. The third dichotomy appeared when we were talking about the inside of the school as unreal and the outside as real. In this phase of the process, we were not able to put off the yoke of dichotomies.

Although we did not reach beyond our prejudices concerning our view of human bodies or dichotomous notions, there were some moments of intra-action when we ventured beyond the human-non-human division and beyond taken for granted beliefs in other senses. For instance, in that moment when we completely disconnected from trying to decipher what the student or the teacher might have meant by his or her statements. At those moments it was just the text itself that entangled with us in the intra-activity. In our interpretation of what diffractive reading entails, we believe that we have experienced what Hultman and Lenz Taguchi (2010) describe as becoming-with the data. This means that we have experienced bodily affective learning, a knowing that emerges in-between (Hultman and Lenz Taguchi 2010, p. 538).

12.8 Ethical Dimensions

We can draw a conclusion that our diffractive reading has contributed to the understanding of our excerpt in a different way. The knowledge we have gained about our diffractive reading in the first place was about ourselves, our material bodies and the prejudices towards images of other material bodies. Although we normally try to avoid work being influenced by bias, our prejudices are still there and affect our understanding of the excerpts. This time we allowed prejudices to have a voice, and in the first reading, this was clear, since our prejudices were brought out of the closet and became apparent. Being in the intra-action with the texts, we were affected by the materiality, the non-human and the excerpts. The words affected us to give way for our prejudices. The trickster in our bodies also reacted. One might say that we got hold of diffractive reading saying yes to a gaze from within, in contrast to a view from the above, i.e. "…a view from a body, always a complex, contradictory, structuring, and structured body versus the view from above, from nowhere, from simplicity" (Haraway 1991, p. 589).

This is the way of doing partial and situated research, the only way of doing objective research as Haraway (1991) and Barad (2003) see it. Positioning ourselves as researchers in this place, in-between, means taking ethical responsibility when seeing the object as equally active and meaning-making as ourselves. We believe that revealing our prejudices is an ethical gesture. This opening of "Pandora's box" might be seen as giving up our control, to be influenced by the non-human and an uncontrollable trickster (Haraway 1991). Hence, by reading diffractively means to go all the way down the lane towards what Haraway and Barad regard as partial and situated research, which means to be in the world and not outside and to regard all matter having agency.

In the reading, we ascertained another perspective compared to earlier work with our excerpts, a general perspective instead of getting lost in the actual classroom or student text. It clarified that we cannot disconnect our bodies from the materiality. We are part of the process and thereby involved in all parts and thus influence research, which tend to be something that we normally hide. Actually, we were produced by texts in the form of data, theory and analysis that act with a material force; material forces are not the sole author nor are we of such material texts (Mazzei 2013). Thereby we have taken an ethical responsibility in this transparency as well as when seeing the object as equally active and meaning-making as ourselves. Emotions have been brought to the surface and have been used, since we cannot pretend that we are a mind without a body as the Cartesian thought stipulates and which natural science is still leaning on.

We claim that using the diffractive reading tool from a material feminist perspective has given us insight that would not have been possible when taking the anthropocentric and positivistic perspective. It has also proven to be a more upright, and thereby more ethical, alternative compared to the positivistic versions when studying science education. We have seen a better scientific description of the world, *a view from somewhere* that Haraway asks for (1991). We have recognised our prejudices and exposed them instead of trying to pretend that they do not exist. What Haraway (1994) calls *the background* in the string figure game has come out of the cupboard and is shown to be alive. Our research has thereby given a description on how entwined and inseparable the cultural, social, political and material perspectives are in a research study.

References

Alaimo, S. (2008). Trans-corporeal feminisms and the ethical space of nature. In S. Alaimo & S. Heckman (Eds.), *Material feminism*. Bloomington/Indianapolis: Indiana University Press.

Barad, K. (2003). Performativity: Toward an understanding of how matter comes to matter. *Signs, 28*(3), 801–831.

Barad, K. (2007). *Meeting the universe halfway. Quantum physics and the entanglement of matter and meaning*. Durham/London: Duke University Press.

Brickhouse, N. W. (2001). Embodying science: A feminist perspective on learning. *Journal of Research in Science Teaching, 38*, 282–295.

Deleuze, G. (1990). *The logic of sense*. Trans. M. Lester. New York: Columbia University Press.

Halliday, M. A. K. (1996). *Writing science. Literacy and discursive power* (Vol. 5). London/New York: Farmer Press.

Haraway, D. (1988). Situated knowledges: The science question in feminism and the privilege of partial perspective. *Feminist Studies, 14*(3), 575–599.

Haraway, D. (1989). *Primate visions. Gender, race, and nature in the world of modern science*. New York/London: Routledge.

Haraway, D. (1991). *Simians, cyborgs and woman. The reinvention of nature*. London: Free Association Books Ltd.

Haraway, D. (1994). A game of Cat's cradle: Science studies, feminist theory, cultural studies. *Configurations, 2*(1), 59–71.

Haraway, D. (1997). *Modets_Witness@Second_Milennium.FemaleMan©_Meets_OncoMouse™*. New York/London: Routledge.

Haraway, D. (2008). Otherworldly conversations, Terran topics, local terms. In S. Alaimo & S. Heckman (Eds.), *Material Feminism* (pp. 157–187). Bloomington/Indianapolis: Indiana University Press.

Hultman, K., & Lenz Taguchi, H. (2010). Challenging anthropocentric analysis of visual data: A relational materialist methodological approach to educational research. *International Journal of Qualitative Studies in Education, 23*(5), 525–542.

Jackson, A. Y., & Mazzei, L. (2012). *Thinking with theory in qualitative research: Viewing data across multiple perspectives*. London: Routledge.

Lenz Taguchi, H. (2012). A diffractive and Deleuzian approach to analyzing interview data. *Feminist Theory, 13*(3), 265–281.

Lykke, N. (2009). *Genusforskning: En guide till feministisk teori, metodologi och skrift.* [Gender studies: A guide to feminist theory, methodology and writing. Own trans.] Stockholm: Liber.

Mazzei, L. (2013). Materialist mappings of knowing in being: Researchers constituted in the production of knowledge. *Gender and Education, 25*(6), 776–785.

Milne, C., & Rubin, K. (2011). Embodying emotions: Making transactions explicit in science education. *Cultural Studies of Science Education, 6*, 625–633. doi:10.1007/s11422-011-93554-2.

Östman, L. (1995). *Socialisation och mening. No-utbildning som politiskt och miljömoraliskt problem* [Socialization and meaning. Science education as a political and environmental-ethical problem.] Doctoral thesis, Uppsala University, Uppsala, Sweden.

Svennbeck, M. (2004). *Omsorg om naturen. Om NO-utbildningens relativa traditioner med fokus på miljöfostran och genus* [Care for nature. About the selective traditions in science education, with a focus on environmental education and gender.] Doctoral thesis, Uppsala University, Uppsala, Sweden.

Taylor, C. A., & Ivinson, G. (2013). Material feminisms: New directions for education. *Gender and Education, 25*(6), 665–670.

Chapter 13
Troubling Norms and Values in Science Teaching Through Students' Subject Positions Using Feminist Figurations

Marie Ståhl

13.1 Background

Several researchers argue that the norms and values of science teaching in school are based on its culture and traditions, affecting students' interest in the subject and their view of nature, society and people (Brickhouse 2001; Östman 1995, 1998; Roberts 1998). Within feminist research, many have long since pointed out science as an exclusionary culture with androcentric features (Fox Keller 1977; Haraway 1988; Harding 1986). They argue that it is a relic since the late 1600s when modern science emerged and women were excluded. Accordingly, Cartesian rationalism was adopted, and its dualistic tradition of thought divides life into, among others, male/female, culture-nature and mind-body. These dualisms have survived to this day in science class, textbooks and tests (see Ah-King 2013; Berge and Widding 2006; Ståhl and Hussenius 2016; von Wright 1999). In school science practice, they appear in the form of an objective, logical and rational appearance, for example, the science language in textbooks (Halliday and Martin 1993), and in the way students are expected to think, speak and act (Alsop 2011; Arvola Orlander and Wickman 2011; Ståhl and Hussenius 2016). Hence, it is not surprising that students often perceive this practice as very special and strange, as free of values and subjective judgments and, at worst, as something that does not concern them (Lemke 1990; Östman 1995; Sjøberg and Schreiner 2010). This belief may be enhanced when the myth of the solitary male genius such as Einstein is reproduced along with its stereotypical contrast, often incompetent, or women made invisible. Ultimately, it creates the image of science as being difficult and inaccessible, only provided for an elite (Lemke 1990, 2011).

M. Ståhl (✉)
University of Gothenburg, Gothenburg, Sweden
e-mail: marie.stahl@gu.se

K. Otrel-Cass et al. (eds.), *Cultural, Social, and Political Perspectives in Science Education*, Cultural Studies of Science Education 15,
DOI 10.1007/978-3-319-61191-4_13

Nancy Brickhouse argues that, if we are to create a scientific education with which students can identify, we need knowledge that is based on "… how students engage in science and how this is related to who they are and who they want to be" (Brickhouse 2001, p. 286). From a broader perspective, this means saying yes to an embodied science (Arvola Orlander and Wickman 2011; Brickhouse 2001; Haraway 1991), in which the whole individual contributes to creating science, unlike a science that separates the mind and body. She argues, supported by Etienne Wenger (1998) theories of learning, that we ought to find out what students are involved in and interest them instead of doing the other way around as many previous projects have tried to do, in various ways that attempt to attract students to science. This should be done in relation to what the students identify with and want to identify with because "Learning is not merely a matter of acquiring knowledge, it is a matter of deciding what kind of person you are and want to be … " (Brickhouse 2001, p. 286). The current study is part of a thesis in which this perspective is partly the focus. In a previous study (Ståhl and Hussenius 2016), the norms and values in the Swedish national examinations in chemistry for the period of 2009–2012 were examined from a feminist and critical didactic perspective. The results show that the Cartesian and androcentric features in science as Evelyn Fox Keller (1977), Sandra Harding (1986) and Donna Haraway (1988) revealed very long ago also appear in the Swedish national tests in chemistry.

13.2 Aim and Research Questions

In the present study, the aim is to problematise the androcentric and Cartesian characteristics of the Swedish national tests in chemistry in relation to how students relate to these norms and values. The aim is also to show a possible alternative chemistry teaching with the help of feminist figuration theory (Braidotti 2002, 2011; Haraway 1991, 1997). Teaching encompasses everything that students encounter in the chemistry classroom and thereby also the national tests.

A total of 188 written student responses from the 2009 Swedish national tests in chemistry were investigated to obtain views on how students relate to the norms and values in the chemistry test. The first part was designed to examine the positions that students take in their reply in relation to issues used in the test item and to science, technology, nature, society and people. The second part interpreted the students' positions in terms of feminist figurations. Feminist figurations should be seen as alternative interpretations of students' attitudes/criticism to the actual discourse in the test and as a positive vision and alternative to this discourse. The critical side of the figurations consists of what students express in their texts, which opposes the prevailing discourse of the tests in terms of both textual content and linguistic expressions. At the same time, this critical side is seen as positive visions for an alternative to the hegemonic and androcentric chemistry that the tests mediate. From these positive visions, I have aimed to sketch a possible chemistry teaching based on critical didactic and feminist perspectives. In relation to these aims, the following questions have been formulated:

- How do the student texts relate to the prevailing discourse in the test, that is, to science, technology, nature, society and people?
- What subject positions appear in the texts based on the answers originated from the above question?
- What possible figurations emerge based on the different positions formulated?
- How can a sketch of science teaching be formulated, based on the visions of the feminist figurations?

13.3 Theoretical Background

The theoretical framework for this study aims to understand and gain perspective on the androcentric norms and values in the Swedish national tests in chemistry that are mediated to students. The framework consists of a critical didactic perspective/curriculum studies (Englund 1997; Klafki 1997; Roberts and Östman 1998) and a feminist perspective on the culture, teaching and learning of the natural sciences (Brickhouse 2001; Haraway 1988, 1991; Harding 1986; Miller 2006). With this framework, the assumption is made that there are norms and values in all teaching, the so-called offer of meaning (Englund 1997) or companion meanings (Roberts and Östman 1998), and they can be androcentric. These norms and values, mediated in teaching, are developed based on the interpretations of curriculum and syllabus, subject traditions and teachers' own aims and objectives in order to form good teaching. Accordingly, they somewhat reveal which scientific traditions, view of knowledge, nature, society and people underlie the design of the national tests. The assumption includes viewing these norms and values as affecting students' interest in a subject as well as their view on nature, society and people (Englund 1997; Miller 2006; Östman 1995).

13.4 Interpretive Framework

In this study, Judith Butler's performativity theory (2007, 2011) is used to understand how the test calls on students to embrace a "scientific position". In Butler's theory, the focus is on the creation of a subject position in terms of gender. In this study, the perspective is broader and more general and also applies to other subject positions that students express. A materialistic vision is assumed thereby that the whole context is materialised before the students, including the students themselves who have their own agency (Barad 2007; Haraway 1991, 1997). The students' text is viewed as embodied, meaning that the actors/subjects have thought and acted through their bodies, with all their physical abilities (Braidotti 2002; Brickhouse

2001; Haraway 1991). Butler argues that gender is performative, and gender identity is created through repeated acts. When repeated enough times, these acts solidify and become fixed in a special form. A norm for a particular identity is created, for example, how one should look and dress and what characteristics one possesses, properties which become biologically essential (Butler 2007, 2011). This study assumes that the test and any previous teaching of chemistry have "called on" (Butler 2011, p. 81–82) students to adopt a scientific identity that tells them how to behave as scientists, speak and think and, among other things, how to relate to nature, society and people. These discourse borders can be seen not only as obstacles to students but also as a means to producing the right products from a teacher's perspective, in this case, to produce future scientists (Foucault 2008).

I am also inspired by Dorte Marie Søndergaard's (1996) description of the gendered body as a gender-marked body that, in different contexts and discourses, is attributed to different associations. Possible associations that students in this study may form are that chemistry/science is difficult, inaccessible and elitist and that scientific work only takes place in the brain (Lemke 1990). It may also produce associations that scientists are not women since, in the 2009 test, they appear to be clumsy, irrational and emotional in both words and images and in contrast to the merited image of how scientists should act (Ståhl and Hussenius 2016). The androcentric associations may have also been strengthened by the use of the oil and mining industry and its products in the majority of the test items, since these occupations traditionally have a male gender structure.

13.4.1 Feminist Figurations

The concept of feminist figurations is described by Nina Lykke (2009 p. 49) as a vision of a different and positive way of looking at the subject that breaks with the androcentric hegemonies by which society is characterised, in which the norm is a man, and where male and female subjects are often regarded as their opposites: good versus less good. Science practices can be seen as androcentric hegemonies (see Haraway 1988; Harding 1986; Hasse 2002), which the previous study (Ståhl and Hussenius 2016) of the Swedish national test in chemistry (2009–2012) has also proven to be. Above all, it is about the remaining positivist view of science as objective, rational and logical, which in the inception of modern science in the 1600s was attributed to male characteristics contrary to women's and nature's emotional and irrational behaviour.

If Butler's performativity theory and Søndergaard's association theory help to understand how the subject positions are created, Rosi Braidotti's (2011) and Haraway's (1991) theories of feminist figurations (Lykke 2009, p. 48) reveal that the nomadic subjects and the cyborg are the basis for these study's views on the subject and also the analytical tool of students' subject positions. The term "subject" refers to mind, body and emotions, everything that makes individuals human. In Braidotti (2002, 2011) and Haraway's (1991) interpretation, the subject is creative. It has its own agency, can offer resistance and can change its situation while drawing inspiration from its past experiences. From this perspective, the subject always has the option to choose which position to take. According to Haraway (1991), humans are all cyborgs, hybrids of machine and organism. The cyborg is a biocultural process in constant change with its fluid boundaries that, among other things, is challenging the Cartesian dichotomies of nature and culture, body and mind. Haraway (1991) argues that, with the help of the cyborg theory, many of the dualisms that arose during the Age of Enlightenment can be fought. In Haraway's cyborg manifesto (1991), the human body is united with its gender and can be understood "... as well as discourse, as the result of human and non-human actors' interaction and trickster ... dimensions that cannot be separated" (Lykke 2009, p. 98, translated by the author from Swedish). A trickster is an intrinsic agent, present in all matter, and can be interpreted as an expression of a behaviour counter to a prevailing discourse or against what is considered right and proper. In other words, the cyborg figuration can help determine whether and how students in their answers are challenging the scientific dichotomies and how they relate to the human dependency on science and technology.

Haraway's cyborg and Braidotti's nomadic figuration also entail that the subject is entirely bodily present and acting with its own agency, contrary to the Cartesian thought where only the mind is present and active. Hence, the subject in Braidotti's and Haraway's theoretical framework may offer resistance and change its situation. The figuration concept proposes that students can deal with the conditions and terms that apply to the here and now as well as write answers to test items in chemistry. It also signifies a movement towards something else, "... this nomadic style is an integral component of the concept of 'becoming' ..." (Braidotti 2002, p. 8). According to Braidotti (2002) and Haraway (1991), it can be regarded as a force that is driven forward in the form of other possible alternatives, for example, to think, act and offer resistance as well as provide constant feedback to what is currently happening. Like Braidotti, I compare this concept to embodiment "... to think through the body, not in a flight away from it. This in turn implies confronting boundaries and limitations" (Braidotti 2002, p. 5).

The figurations presented in the second phase of this work, which are alternative interpretations of students' subject positions, should be regarded as fictional and potential but not as being fixed and finished.

13.4.2 Apparatus of Bodily Productions

Inspired by Haraway and the "apparatus of bodily production" (Haraway 1991), I want to emulate the test at an apparatus, which emerges from the interweaving of various "material-discursive" (Alaimo and Hekman 2008) practices and in the meeting with the students, which give rise to different positions. Thus, the meeting between the students and the test is akin to what the Science and Technology Studies (STS) and what Bruno Latour (1999), Karin Knorr Cetina (1999) and Andrew

Pickering (1995) describe as an interaction, a mangling. In their case, it is between "apparatus" (the researcher's object of study/tool) and the researcher.

Both the apparatus and humans have agency as well as hold an uncontrollable trickster (Haraway 1991). The subject positions should thus be seen as discursively constructed and "mangled" by the joint interaction between the apparatus (the test) and students and both their inherent and uncontrollable tricksters. The mangling can be described by Butler's performativity theory (2007, 2011) that the test calls on students to take a scientific position, for example, be objective, logical and rational, that students react to this and that they act unconsciously through their tricksters (Haraway 1991). The trickster in the test may be such that the test designers are not aware of it, unlike the conscious motives behind the test items. Previous research (Ståhl and Hussenius 2016) has noted this issue, the so-called offers of meaning (Englund 1997), norms and values found in the tests, in the form of, for example, gender bias. The student trickster can be students' unconscious reaction to androcentric and anthropocentric features in the test or other excluding traits. Together, this mangling between the test and the students produces different positions.

13.5 Method

This study examines a specific discourse, the 2009 national tests in chemistry and related student responses, with its particular norms and values to which students should relate. How students relate to this practice is first examined using a discourse analysis to determine the possible positions within which students position themselves. Subsequently, the subject positions have been interpreted on the basis of Haraway's (1991) and Braidotti's (2011) feminist figurations.

The discourse analysis is inspired by James Paul Gee's model of discursive building blocks (Gee 2014). This includes viewing the students' verbal statements as seven actions that together build something. In this case, I study how students build subject positions in relation to the scientific position that the test is mediating. These consist of the following building blocks in the analysis requested in the students' texts:

1. What is highlighted in students' texts (e.g. science and technology achievements or its negative sides)?
2. What connections to other topics are made in relation to the subject matter of the test (e.g. hygiene, cooking, medicine, nature, which the test does not include)?
3. What is valued explicitly or implicitly in the texts (technology, nature, human behaviour, people's bodies, other people, the world, things)?
4. To what extent and in what ways are values and reinforcing expressions used, that is, how high is the evaluative text volume in the texts (Ståhl and Folkeryd 2016)?
5. What language system is used (e.g. everyday language or scientific language)?

6. How is the relation to the reader presented in the students' texts, (are personal pronouns or passive form used)?
7. Which subject positions are projected on the basis of the answers that have emerged to the above questions? Do students stick to the stipulated discourse or do they choose to deviate from it? If the latter occurs, which subject positions do they assume?

The subject positions must be seen in relation to the scientific position that the chemistry test mediates its views on science, nature, society and people (see Ståhl and Hussenius 2016).

In a previous study (Ståhl and Folkeryd 2016), the analytical tool appraisal was used to study the positive and negative evaluative expression in students' texts in relation to the topics about which students have chosen to write. Together, these linguistic expressions constitute the text prosody, here called *the evaluative text volume*. Prosody refers to the rhythm, dynamics and stress in a text and can therefore reveal the writer's commitment and interests (Folkeryd 2006). The analysis was meant to assess the degree of evaluative text volume in student texts. In the current study, these results are used to categorise the subject positions that students present. The focus is to what extent and in what ways values are projected in the texts and thus the degree of evaluative text volume the texts convey. This means that the fourth building block can be answered by the results of this previous study (Ståhl and Folkeryd 2016).

13.6 Empirics

The empirics in this study include the 2009 Swedish national tests in chemistry for 15-year-olds, the assessment guidelines and 186 student responses from one of the test items in the chemistry test. Each year, schools randomly collect a number of student responses. In 2009, a total of 200 responses were gathered out of approximately 33,000 possible responses, from students born on certain dates of the year. Each year a total of 100.000 students take the national test in chemistry, physics and biology. Therefore, one-third of the students take the chemistry test. The Swedish National Agency decides which schools should take which test. This means that students do not know in advance which of the three tests they will have to complete.

Of the 200 collected student responses, 186 have been used to answer this survey. The student responses excluded lacked answers to this study's specific item alternatively applied to summary to adequately be analysed. The test item (Fig. 13.1) from which the students' responses are derived is an open task: there is not only one right answer but many, and students are able to express themselves relatively freely. The item, as shown below, relates to the social, ecological and economic issues based on the Education for Sustainable Development (Government Official Reports: SOU 2004, p. 103). The aim of the item from the Swedish National Agency for Education according to the guidelines is to test students' ability to provide examples of industrial applications in chemistry and give examples of how the oil refining industry

Products derived from crude oil have the last 100 years changed the way we live. Crude oil is a mixture of different hydrocarbons. By refining, cracking and other processes crude oil converts into various products.

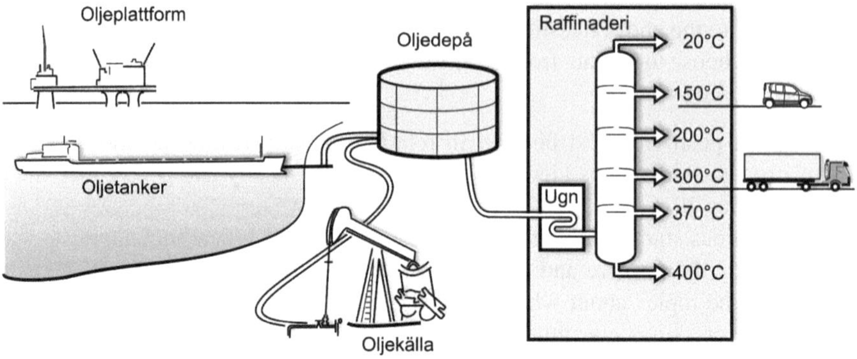

Petrol and diesel are examples of products made of crude oil.

a. Give another example of a product made from crude oil that has affected how we perceive the world around us.
b. Explain how the discovery of refining crude oil has had an impact on how we live and how we perceive the world around us.
c. What made us start refining crude oil?
d. Choose another discovery in chemistry. Describe how this discovery has changed humans' ways of thinking and doing things.

Fig. 13.1 Test item from the 2009 assessment (Skolverket 2014) (Courtesy of the Swedish National Agency for Education)

and other selectable discoveries in chemistry have affected people's lives, perceptions of the world and way of thinking. It consists of four sub-items (a–d) of which the first three concerns (a–c) oil refining, whereas the last item (d) provides an opportunity for students to choose a discovery in chemistry to discuss.

Given that students often make no distinction between the first three sub-items dealing with oil refining but write about them all in the same context, I have chosen to deal with these as one item. In the future, the sub-items in those cases where they are mentioned will be specified as (a, b, c) and (d).

13.7 Results and Descriptions of the Derived Subject Positions

Based on the chemistry position that the test enhances, students are advised to adopt a specific scientific perspective from which to reason and view the world. This includes using only the rational, logical and objective intellect and is thus distinct from the emotional, irrational and subjective expression that the body produces. It also involves viewing the world and nature from an anthropocentric Western standpoint (Ståhl and Hussenius 2016).

The results of the completed discourse analysis contain three main categories based on students' outspoken or non-declared relation to science and technology. In their texts, some students take a (1) *scientific positive position*, an intermediate category where students can be said to stand on their own two feet, a (2) *scientific positive and critical position* and a third variant in which students remain quite a distance from science and technical inventions and a (3) *scientific critical position*. Group 2 means that the students, on the one hand, affirm the scientific achievements. On the other hand, the students criticise them. Based on this overall positioning, several subgroups have emerged. Below is a description of each of the overall discourses and their subgroups.

1. Description of the scientific positive position (71 texts)

Common feature of the scientific positive position texts is that they provide a thoroughly positive overview of the scientifically and technically produced society. This is accomplished although to varying extents. The majority of the texts have a low evaluative text volume. These texts often only implicitly express a positive and personal hold position, although this is possible and permissible on the basis of how the item is formulated. However, a small group of texts is exactly the opposite. These texts have a high or very high evaluative text volume and express many explicit and positive values about discoveries/inventions in science and technology. Criticism towards crude oil, its products or how it is used does not appear in any of the texts. Thus, environmental issues are not part of the students' reviews, and nature is missing. People are rarely mentioned or appear only very peripherally. One could state that the texts are "tinglified" and that the real agent (the human) is hidden. This is a known feature of scientific texts. However, in such a context, this is accomplished by using the passive tense and by turning verbs into nouns (Halliday and Martin 1993). In students' texts humans are omitted, and these grammatical traits are very rarely employed.

All students whose texts concern the scientific positive position have answered the first part of the item (a, b, c), the one about how the oil refining has influenced society. They all responded by enumerating what "we" have received in terms of an increased standard of living and principally in the form of time gains thanks to oil refining. The question of how "we perceive the world around us" is either not answered or answered by stating that the world is perceived as being smaller because travel time has been reduced. Most texts belonging to the positive position may be described as stationary. The texts indicate that students feel satisfied with society regarding their standard of living and they rely on technical and scientific solutions if problems arise.

Given the significant discrepancy in the evaluative text volume among students' texts, which in turn brings other differences, there are two subgroups of this position: (1a) *the reserved positive position* and (1b) *the enthusiastic position*. Below are examples of texts and descriptions of these two positions.

(1a) The reserved positive position

Example 1
(a, b, c) Plastic. We drive car powered by gasoline and call with plastic cell phones.

(d) The discovery that atoms can be split has changed many people's thinking. Einstein's theories have also changed many people's thoughts. The development as a result of the things above is the reason that we nowadays have nuclear power plants and that we can extract an incredible amount of energy.

Example 2
(a, b, c) Ethanol is another product made from crude oil, which has affected what fuel we use for cars and other motorised vehicles. Refining has influenced us because we have gasoline and diesel which means that our motorised vehicles can be driven. We wanted to explore what we could produce from crude oil.

(d) A different chemical discovery is glycol which we use in the cooler in cars and other motor vehicles to prevent the water from freezing.

The two texts above show a very low evaluative text volume. When evaluations appear, they are usually implicit. A positive attitude is, in other words, not explicitly expressed. For instance, comments on how nuclear plants can extract an incredible amount of energy that we can operate motorised vehicles using petrol and diesel and that glycol prevents water from freezing have been considered as implicitly positive evaluations. The above and all other texts within the group comply with a kind of template, that is, to an enunciation and statement of what we can do, thanks to things such as crude oil. In isolated cases, explanations are also provided as to why this is good. For example, Example 2 expresses that both gasoline and diesel are good for driving the vehicle and glycol is good because otherwise the water in a car's radiator would freeze.

(1b) The enthusiastic position

Example 3
(a, b, c) Diesel and kerosene. With the help of crude oil, one can do a lot of things. One can make engines work, heat houses, lubricate and make clothes. All large engines are run by some kind of petroleum products. And without machines today, we would hardly be able to survive. But we can also make polyester fabric which is made of oil. Sometimes, we even burn oil to heat houses. When trying to invent an engine, one needed something to make it work. So, it was probably somebody who understood that crude oil is easily ignited and used as an ignition gas in the engine. We also used petroleum products in paraffin lamps, but they have been replaced with oil.

(d) To create and use electricity. Electricity is the most widely used on earth. Without electricity, we would die. We use electricity to do almost everything. Heat, light, cars, tools. Electricity is generated by turbines that lead to a reactor that creates electricity. Electricity is produced by air, water, fire and nuclear power.

Example 4
(a, b, c) Asphalt from crude oil has made a big difference for humans. We would not be able to drive as easy and quickly if we did not have asphalt roads. If we had not been able to refine crude oil, we could not produce different kinds of oils, petrol or

diesel. Then, we might not have been able to drive or fly anywhere. We humans were a little more settled in our own countries then. For example, it would not be possible to travel abroad for 1 week only. We began refining crude oil because we knew one could extract a variety of things from it. We also understood that the lightest hydrocarbons would evaporate first when heated, so refining crude oil was a good option.

(d) Nitroglycerine. If explosives had never been discovered, tunnels and foundations would have to be excavated by hand. It has been important to be able to build better roads and enable people to build their houses where they want. Being able to blast rocks has resulted in more opportunities to do things. For instance, make caves and such.

Unlike the *reserved positive* texts, the *enthusiastic positive* texts make explicitly positive evaluations towards science and technology. These are emphasised through discussing our high standard of living and set against its opposite, a life without oil and technological achievements, a life in the Middle Ages using horse and carriages or the end of civilisation. Example 3 provides a good example of this: ... *without machines today, we would hardly be able to survive*, or ...*without electricity, we would die.* The discovery of oil and other technical scientific solutions are, in these texts, considered vital. With the help of these contrasts and the accentuated positive attitude, the lack of criticism becomes even clearer, and natural science is shown to be infallible. The texts show a strong faith in science and technology's ability to solve all the problems that society is facing. Despite this, problems are rarely presented in the texts. They only appear in situations concerning our bodies.

Example 5

(d) How to purify water. If we had not discovered this, we would not be able to drink water or be so healthy today.

Example 6

(d) All medicine currently used is obtained chemically. Without these drugs there had been many deaths today. Many mentally ill individuals would never be able to recover.

That water can be dirty and make humans become sick is formulated in the texts as obvious problems from which science can save us. Our bodies are vulnerable, but not science. These examples are, from the scientific positive position, among the few texts that mention people and the human body. Four other texts implicitly mention the human body, but in more rational terms, as in the example below.

Example 7

(d) Radiation. Radiology and diagnostics improved, and they started to think about internal injuries.

Among the enthusiastic positive texts, all students answered item d; this is only found in one-third of the reserved positive texts. The majority of the enthusiastic positive texts mention inventions/discoveries that are characterised as an important part of our daily lives and are close to the human body. They are also characterised by the choice of topics to discuss that do not conform to the test choice topics. They include medical discoveries, hygiene products, packaging and storage innovations, polyester fabric, photography and light bulbs. Of these, six have chosen to talk about elements associated with the needs of the human body in the form of soap, shampoo, food, medicine and clean water. Regarding the reserved positive texts, the one-third that has answered item d has focused on petroleum products or car and oil products. Alternatively, they have chosen something that was already included in the test (metals and dynamite) or inventions in the energy sector, nuclear power plants and electricity. These students can be said to follow the tests' chosen path of subject matter, whereas the respondents of the enthusiastic positive texts have taken a different path based on everyday life.

2. Description of the positive scientific and critical position (89 texts)

In the positive scientific and critical position, the students are expressing positive values of our high standard of living that we enjoy thanks to oil refining and other chemical discoveries. At the same time, they are not satisfied with the status quo. Unlike the positions taken by the tests and the scientific positive position that both regard discoveries and inventions from a positive perspective, the positive scientific and critical position also considers the downside.

All texts in the scientific positive and critical position hence express positive evaluations about scientific and technological discoveries/inventions as well as criticism of the same. The criticism is directed both against the oil industry's impact on the environment and against its links to armed conflicts involving the atomic bomb and people's amoral behaviour in terms of issues such as environmental awareness, consumption and abuse. Since the texts sometimes treat one or more of these critical inputs, the overall position is divided into the following subgroups: (2a) *the scientific positive and environmentally critical*, (2b) *the scientific positive and environmental and ethical and/or moral critical* and (2c) *the scientific positive and morally critical*. Below is a description of each perspective. The middle position is also divided into ethical and moral criticism, (2b.1) and (2b.2).

(2a) The scientific positive and environmentally critical position

Example 8
(a, b, c) Asphalt is used on roads for cars, bicycles, buses, etc. We walk or drive on it every day. It was positive to start refining crude oil from an energy perspective, but it's not so good for the environment. We use oil of some type every day, such as fuel and asphalt, but the environment is damaged by it. Fish die, animals get sick and plants die, but they have a little hope. Since it takes millions of years to produce new crude oil, it will one day run out.

The writer believes that it was beneficial from an energy perspective to refine oil but not for the environment. The criticism is of human's use of oil and its impact on the environment. Nature is tangible and concrete. Animals and plants are referred to and described according to their fragility; they can get sick and die as a result of our daily use of oil products. However, the student does not moralise about human behaviour. The problem will be solved when the oil runs out.

(2.b.1) The scientific positive and environmental and ethical critical position

Example 9

(a, b, c) It's mainly gasoline and diesel fuel that have affected us both positively and negatively. Gasoline enables us to drive. We can visit places faster. If oil had not been produced, it would take me 3 hours a day to get to school; now it takes 25 minutes. But oil has also affected us negatively; only to produce oil is hazardous to the environment and all ecosystems and recycling. Not only nature has been adversely affected, but also oil also creates wars around the world, such as in Iraq and Saudi Arabia where lots of people have died because of it.

In this text, oil's impact on the environment is criticised by how it leads to armed conflict, but humans are free of guilt. It is oil's fault that the environment has suffered and that many people have died. It can be seen as a tinglification that the blame is transferred to non-humans. This can be compared to how written natural science language objectifies and hides the "real" agent through using the passive tense. By using the word "people", a certain distance to those who die is also described. The writer himself/herself is, however, clear about describing the positive time gains obtained through the use of oil products. Even nature is described concretely by speaking of ecosystems and natural cycles, and the term "nature" is used in contrast to the "environment". This view of nature inclines to biocentrism. Thus, it involves seeing nature's well-being based on nature's own sake and not anthropocentrically. In this text, countries considered to be outside of "our world" to which the test item is referring are highlighted. A major topic of conversation in student texts, which is also visible in the above text, is weapon production in the form of inventions such as the atomic bomb, nitroglycerine and residual products from nuclear power plants, which have created an unstable world where many people have been killed. When students in test item d may choose which discovery/invention to discuss, many write about weapon production, as the following text shows.

Example 10

(d) I think nuclear weapons (atomic bombs) have changed the way we think the most. It was quite an unnecessary discovery used for nothing but evil. Therefore, a lot of people have died in wars and so on, thanks to atomic bombs. And it has not been used for something good, and it will certainly never be. It has made us humans more aggressive and callous, I think. Not that war is ever good, but it was better in the past when one fought with horses and swords instead of guns, aircraft and bombs.

The above text shows that the writer has her/his own agency who, in contrast to the national test, is not hiding the ethical and emotional dilemmas that the use of oil and other natural science discoveries/inventions brings with it. In the above text, it is mainly nuclear weapons that are condemned. The evaluations are clearly emotional, as opposed to the objective and rational language mediated by the chemistry position in the test. The text has a high evaluative text volume, and the student can thus be considered to have a high commitment to what she/he writes about. Additionally, by choosing to mention nuclear weapons, the student expands the discussion beyond the topics that the chemistry test has chosen to present, which focus on the mining and oil industry (Ståhl and Hussenius 2016). In the above text, the student sheds light on an invention that has negative overtones. The test does not highlight the fact that there are negative aspects of discoveries and inventions in chemistry.

(2.b.2) The scientific positive and environmentally and morally critical position

In many texts human behaviour is singled out explicitly and turned into a moral issue. More than half of the students have focused on negative human behaviour in relation to environmental problems, armed conflicts, drugs, alcohol and body ideals. One could state that it is partly akin to the traditional deadly sins in the form of human foolishness, stupidity, greed, laziness, selfishness and pride. Some students provide ideas about how moral issues should be resisted. They believe that people do not understand that the earth will perish if oil is not renewed and/or replaced by fossil-free fuels. The message from many students is to cycle more and travel less by car. Sometimes, the criticism is only directed towards others. However, sometimes the students involve themselves in the criticism; in other words they express a … *we have* … instead of … *people need*. The following text is an example of such a moral theme.

Example 11

(a, b, c) Other products produced from refining oil are different varieties of oils and asphalt. When we discovered that we could extract fuel for cars and other vehicles by refining crude oil, we could produce fast vehicles and travel faster, enjoy more comfortable lifestyles and contaminate the soil and drain its resources. We perceive our world as a never-ending resource that gives us everything we want so that we can live as comfortably as possible.

(d) People have discovered various medications through different chemical laboratories. These medicines and vaccines have made fatal diseases almost harmless and raised the average life expectancy in many countries.

This text has a strong focus on human greed and selfishness. We pollute and give ourselves the right to do what we want with the earth's resources for our own convenience. The oil and its products are not valued. However, human's positive side is highlighted when they have succeeded in producing medications that can cure deadly diseases and improve the average life expectancy in many countries. Nature is not spoken of directly, but in terms of the earth's resources. This signals a greater distance from nature than in the two previous texts.

(2c) The scientific positive and morally critical position

The concrete and tangible body is included in the arguments of one-third of the students' texts in the scientific positive and morally critical position. It comes in the form of criticism of people's consumption of fuel, that is, people are lazy and will become obese with such a behaviour. It is also shown in students' answers to sub-item d where discoveries and inventions that concern the body, such as medicines, hygiene products, contraceptive pills, radiology, anatomy, drugs, alcohol, carbohydrates, fats and proteins, are discussed. Below is such an example.

Example 12

(a, b, c) One example is plastic that we use in everything from clothing to furniture. To refine oil means that one can separate the various hydrocarbons from each other. Various hydrocarbons burn at different rates, and that means they are useful in different situations. The discovery of refining has enabled us to use oil for a tremendous breadth of things such as gasoline, cosmetics and paints. This affects our way of living in all areas where we use products derived from refined oil.

(d) One discovery is that polyunsaturated fats are healthier than single saturated fat (which scientists currently claim). This is one of several things that have made people in today's society more aware of how their body is affected by the food they eat and more able to make wise decisions. Now that obesity, cardiovascular diseases and unhealthy people are more common than ever, there is a hysteria about lowering cholesterol, losing weight and feeling great. Most food products in shops today include a list indicating how much sugar, proteins, fats, etc. it contains. Under fat, it states which are saturated and unsaturated, sometimes even including polyunsaturated fat. The knowledge that this discovery has given us means that we can change our habits; all we need is to look at the packages first.

In the second part of this text, the body is very real, and lots of "moralising is served". People are not able to make wise decisions, even though the information is right in front of them. This in turn leads to hysteria about cholesterol levels, losing weight and feeling great. The writer talks about people in terms of being hysterical, but she/he uses the pronoun "we" when it comes to making wise decisions. There is seemingly a distance between the writer and the first group, those who misbehave, but a rapprochement with the group taking wise decisions. She/he speaks from a top-down perspective. The student seems to have a split view on science. On one hand, the writer believes that knowledge of, for instance, polyunsaturated fat is important and positive. On the other hand, one cannot trust scientists. They can change their opinions and perhaps even be wrong.

The students, like the above writer, who write about people's negative behaviour, whether it involves unhealthy eating, not exercising or abusing alcohol and drugs, devote only a few words and values to the first three questions, that is, to oil refining. The focus is instead on sub-item d in which the evaluative text volume is high and which the above text expresses.

Summary of the Scientific Positive and Critical Position

A large majority of the texts in this scientific positive and critical position, in addition to the oil theme, concern student's everyday life in the form of a close human body perspective. In these texts, students move between different positions, from the extremes to discussing "for and against". In addition to the close physical issues such as food chemistry content, medicines, drugs and hygiene products, the students talk about the light bulb (to see well), batteries for mobile phones (to have social contact), fire (to cook and warm themselves) and armed conflicts (which lead to death and misery), which are made possible by chemistry inventions through atomic bombs, nitroglycerine and waste from nuclear power. Only seven texts deal with topics that belong to basic chemistry (or physics) which the test has discussed, for example, the structure of atoms, the periodic system, biogas and nuclear power. From a wider perspective, it may be interpreted as students move towards an embodied (Brickhouse 2001) scientific position that does not differentiate mind from the body either from an ethical/moral or emotional perspective. Nonetheless, the majority still maintain quite a strong anthropocentric attitude towards nature. The dichotomy between nature and culture persists in most of these student texts, whereas the one between the body and mind has been torn down, by many students.

3. Description of the scientific critical position (26 texts)

The critical position consists of texts that are entirely critical of scientific discoveries and inventions alternative to human behaviour. This position addresses the same subject areas as the scientific positive and critical position described above. Oil, according to students, tends only to contribute to the environment and humanity by resulting in harm and danger. Half of the students blame this situation on humanity, on our behaviour. We are greedy, selfish and lazy, and we want to live as comfortable as possible. Implicit emotions are manifested in disappointment over the human aversion to understand their own and others' welfare, including nature and animal welfare.

There are three subgroups in this position: (3a) environmental criticism without human intervention, (3b) environmental criticism on the basis of morality and/or criticism of weapon production and (3c) criticism of human morality (implicit environment critic). Below are shown examples of texts from each of these three positions.

(3a) Environmental criticism without human intervention

Texts belonging to environmental criticism without human intervention focus on the environment and nature. The texts briefly describe the consequences of what oil brings to the environment or nature. Usually, oil is blamed. Sub-item d is missing and the textual evaluative volume is low. Regarding the latter, the text below is a rare exception. Albeit short, several strong expressions (underlined) are expressed.

Example 13

(a, b, c) Plastic. Plastic is *unnatural*. It does not decompose. It *breaks down* nature.
(d) The car. *All* exhaust gases are affecting the world *negatively*.

This example is only one of its kind where some reinforcing words/gradations (text in italics) still accentuate the stated values. However, as in the other texts in this position, humans are not involved in the rendered criticism of oil. Humans are not even mentioned. Instead, oil is considered the villain. This attitude towards nature is close to biocentrism since all focus is on how nature is damaged. However, no connection is made that this would be a disadvantage for humans and no solutions pronounced to environmental problems or "hope for better times".

(3b) Environmental criticism on the basis of morality and/or criticism of weapon production

In this group, which consists of ten texts, the message is clear that humans should be blamed for the fact that earth is contaminated. It concerns human immoral behaviour, comfort, gluttony and selfishness. This can be seen in the following text.

Example 14
The discovery of refined oil has made us a little spoiled. It has led us to drive a car instead of walking, and we do not care as much about the earth becoming contaminated as long as we are comfortable.

In the example, there are no extenuating circumstances. However, some blame is put on oil, as it is spoiling us and encouraging us to drive cars and pollute the earth. Nonetheless, we do not care as long as we are comfortable.

Nine of the texts in this position have an environmentally critical focus. Although the tenth text expresses environmental criticism, most of the energy and high textual evaluated volume is placed on alcohol addictive effects in sub-item d, as shown in the following excerpt.

Example 15
(a, b, c) Refining oil has affected the way we live, because we use petrol and diesel every day. It uses a lot of energy and impacts our environment, the climate and the ozone layer.Today, our world needs the benefits of refining crude oil. I wonder how the world would have looked like if we never started doing it.

(d) Alcohol, or, in other words, ethanol, has affected humans a lot, *I think.* Many die because of alcohol. Some have tried to make moonshine (wood alcohol), but instead of becoming ethanol, it may have become methanol. Methanol is really dangerous and has resulted in people's deaths. Some have become addicted to alcohol (ethanol) and drink alcohol every day. They have since become alcoholics. Some alcoholics drink so much that they die. Many young people drink alcohol. Meanwhile, under the influence of alcohol, they are affected by it. One might do things one would never do sober. According to films and texts that we have read, *my understanding is* that people disappear a little from reality. I have never had any need for alcohol, and I *hope I never get it!*

The text begins with an environmental criticism of our use of gasoline and diesel. It is expressed with a low evaluative text volume. In the sub-item d, a black picture is painted of the harmful effects of alcohol. The text becomes much more judgmental and contains a high evaluative text volume. This part shows no strong outspoken criticism of people. Rather, the criticism lies on the alcohol itself. When values are expressed, the writer often writes herself/himself into the text: *...I think,... my understanding is...* The solution in order not to suffer from these negative effects is to avoid entering into an alcohol addiction. The student hopes this will not happen to her/him. This is emphasised by the writer by again putting herself/himself in the text: ... *hope I never get it!* Although the blame is put on the alcohol, the teaching, in the form of films and texts, receives implicit recognition as it apparently helped the writer to abstain from alcohol. This indicates a positive input, but it does not help all people to avoid alcohol.

In a third variant of text within this subgroup position, the criticism is directed against human immoral behaviour in relation to the nature/environment, human relations and human's adverse use of weapons or, as in the text below, towards gunpowder.

Example 16
(a, b, c) They have affected us a great deal. We have become lazier and believe that the oil will never end. We drive to work. We fly to the other side of the globe with an airplane. We get fire by the light push of a button and so on. We believe that the oil will never run out, but it damages nature infinitely. The temperature rises, ice melts, lakes disappear, and more. We must act soon before something terrible happens.

(d) Gunpowder was invented by the Chinese long ago. They used it for rockets, but when West got hold of it, they used it in wars. This has now been developed a lot, which has made humans superior to other animals.

In the text, criticism regarding our laziness and ignorance is addressed to all of us, including the writer herself/himself by the use of the personal pronoun *we*, whereas the misuse of weapons and ammunition is carried out by humans. This is also consistent with the other texts expressing similar moral criticism. That is, the environmental problems are caused by us, whereas armed conflicts, gluttony and greed, among other things, are blamed on someone outside the writer by using, for example, passive tense. The text also has a more global perspective and thus goes beyond the limits of what the test item stipulates "We and our environment". This use of the "us and them perspective" on the test item is furthermore in this text vice versa. The West did wrong when they used gunpowder in war, and man's superiority to animals is not something to brag about when it is achieved by brute force. The author's view of nature can be seen to be considered biocentric. On the other hand, we must act before something bad happens, even though the "lakes and much more" have already disappeared. It can be interpreted that this is not terrible enough: the worst is when it affects us.

(3c) Criticism of human morality (implicit environment critic)

In this position, the environmental issues are implicitly expressed. The focus is instead on people's amoral behaviour and not what that conduct should produce or depend on. The following is an example of this.

Example 17
How this has affected the way we live today is pretty significant. We do not need to think so much because everything is automatic. We do not think about how much everything is destroyed, quickly and automatically. It is cheaper because no one needs to be there all the time and look at it.

As in the text above, the texts in this subposition are mostly short. Nonetheless, several students express clear values (medium evaluative text volume) about our moral compass which is out of play. The text mediates implicitly that, although our ignorance is clear, our unwillingness to see what happens that everything is destroyed and that our fully automated society and aspiration to produce goods as cheaply as possible also means that we do not have to look at it. No direct connections are made to the nature or the environment, but the text can be seen as a general reflection of society's decadence. Human's poor morality is thus clearly in focus.

Summary of the Scientific Critical Position
Generally, the texts in the critical positioning express very little hope of improvement for earth's survival or for people to change their behaviour. There seems to be no way forward. In contrast, the students show that the trickster (Haraway 1991) in nature has reacted and resisted that nature has struck back. Students have also clearly distanced themselves from human foolishness and behaviour, even though they often involve themselves in this human decadence through the use of the personal pronoun "we". There is also no trace in the texts that they rely on science and technology to solve the problems mankind has caused. Students' rejection of science and technology can be said to strike back at the positive image that the chemistry test has painted. Overall, this action can be interpreted as a feminist figuration (Braidotti 2002; Haraway 1991) that has used its agency and showed its resistance towards the science and technology community. This is manifested by the use of emotionally coloured and evaluative language and through highlighting the negative consequences. It is an emotional movement, leading away from science and the path on which the adult world has chosen to walk. Students do not show any way back and do not express any positive views on what science and technological discoveries/inventions have produced. This can be interpreted to mean that students have turned their back on science and technology forever.

13.8 Feminist Figurations and a Possible Common Educational Vision

In the present study, the aim has been to problematise the androcentric and Cartesian characteristics of the Swedish national tests in chemistry in relation to how students relate to these norms and values. The aim has also been to show a possible alternative chemistry teaching with the help of feminist figuration theory (Braidotti 2002, 2011; Haraway 1991, 1997), where other norms and values manifest than the ones that have emerged in the chemistry tests. Through the development of subject positions, student positions have been reported in relation to the norms and values that students have met in the 2009 national test. Thus, one aim of the study has been achieved, highlighting the attitudes students have towards science and technology, to their perception of nature, society and people in relation to the norms and values in the 2009 chemistry test. Subject positions have been obtained in response to the questions I have put to the material. They represent a section of students' positions in relation to the assumed perspective in the research questions. Starting from the developed subject positions, possible feminist figurations have become visible, which in turn can be interpreted as possible visions of the scientific education based on the students' developed written criticism. These figurations are reported in the following sections along with the possible educational visions.

13.8.1 *Feminist Figurations Based on Students' Subject Positions*

The feminist figurations are based on students' subject positions and thereby their attitude and criticism towards the actual situation, in the meeting with the chemistry discourse in the test, and what I have interpreted as positive visions of the subject positions. Thus, I understand the concept of the feminist figuration "… to be understood as a vision that the subject is about to realize, but additionally implies that a critique of the here-and-now- situation. It speaks to thought, feeling and body" (Lykke 2009, p. 220 translated by the author from Swedish). Figurations should be seen as temporary, moveable and partly fictional. They have emerged from a mangling (meeting) between, in this case, students and the test, the apparatus. The latter can be described as "… boundary-making practices that are formative of matter and meaning, productive of and part of the phenomena produced" (Barad 2007, p. 146). The boundaries of the apparatus are, in this study, the norms and values in the chemistry test (Ståhl and Hussenius 2016). Students have had to relate to these discursive boundaries, both with regard to the overall discourse in the chemistry test (2009 national test) and the discourse that applies to the current test item that students have answered.

Braidotti's nomadic subject and Haraway's cyborg can, on one hand, be understood as a new, positive way of doing gender (Lykke 2009). On the other hand, one can also see that Braidotti, with her concept of figuration, also wants to achieve something else: "I rather see them as significant sites for reconfiguring political practice and redefining political subjectivity" (Braidotti 2002, p. 3). Haraway also shifts focus away from the human subject when she states that figurations are "… material-semiotic nodes or knots in which diverse bodies and meanings coshape one another" (Haraway 1989, p. 4). Similarly, I want to see that, in the meeting between students and the chemistry test, nodes/figurations are created that, in relation to the norms and values in the test, take positions for or against science and technology. Therefore, they evaluate people's actions and nature and society, even though they describe a vision for something else. This is based on my interpretation of a possible vision of scientific education that is not based on the Cartesian tradition of thought, reflected on the national chemistry tests. Rather, it is based on students' criticism, ideas, norms and values.

13.8.2 *The Critical Side of Figurations Leads to the Vision of a Different Chemistry Teaching*

Using Braidotti's and Haraway's theory of figuration, each subject position, in addition to being a combined form of several students' positions regarding their relationship to science and technology, nature, society and people, is interpreted as being in possession of an alternative position, a vision of a different chemistry teaching. This means that these students' nomadic subjects/cyborgs can be seen to have exercised their agency and taken a stand against the androcentric hegemony that the chemistry test mediates (Ståhl and Hussenius 2016). It also reflects them striving towards a new vision of chemistry teaching. Some figurations can then be viewed as striking back at the positive image of science and specifically of chemistry that the test has painted. In two of the three subject positions, in *the scientific positive and critical subject's position* and *the scientific critical position*, science adverse consequences for nature, society and humanity have been highlighted. These positions can be regarded completely conflicting with the natural sciences. On the other hand, it can be interpreted as concerning a feminist figuration that says no to the very thing that the test has mediated. Instead, it moves towards a vision, *an imagined elsewhere* (Haraway 1992, p. 295), for a change of discourse. This study takes note of the latter interpretation. Similarly, one can understand the two subpositions of *the positive science position*: the enthusiastic position and the reserved positive position. Students in these groups affirm science and technology and thus seem to lack visions about something else when they do not express any criticism. However, the texts in the enthusiastic position raise a much broader topic than the chemistry test mediates. They move between everything from outer space to everyday duties in the kitchen. This reveal, on the one hand, a feminist figuration who expresses criticism of chemistry of the test-limited choice of topics and, on the other, the vision of a broader chemistry subject that concerns young students. This involves a chemistry teaching that does not get stuck in a structurally masculine choice of topic in the form of oil and metal industry, which has been the case in all four previously studied chemistry tests (Ståhl and Hussenius 2016). Rather, it raises subject matters that are of concern and interest to all students.

In *the reserved positive position*, the students' answers are implicitly positive and characterised by short texts and a low evaluative text volume. Here, a vision is more difficult to see. However, a figuration that breathes "I'm not interested" and a desire to escape from it appears, in other words, an implicit criticism of the test and its contents. The positive vision becomes a science in which everyone gets to be heard and involved. Thus, a science education is built on many students' different visions, including that of human equality.

The science of positive and critical position aims to highlight and make natural science discoveries/inventions from many perspectives, to understand that they are part of a complex reality. Nonetheless, unlike *the positive position*, they also want to show that science and technology bring both the good and the bad, including morally indefensible actions. This is thus a feminist figuration that expresses criticism of the prevailing hegemony that mediates that chemistry is infallible. The positive vision would then be that ethical considerations must be brought to light and discussed in the chemistry classroom. This applies in relation to the problems of society and the individual, which they face/will be facing on both a small and large scale. It is about problems that need to be discussed and clarified, even when they are complex and multidisciplinary.

Science based on everyday and close human phenomena and choice of topics (chemistry in cooking, the body and cell phones, hygiene products and medications) are also discussed in these positions. The body from a chemistry perspective is thus developed. The human side of a cyborg (Haraway 1991) also becomes evident in the students' texts, as the machines we use are from a human perspective and relation. Thus, critics are said to be directed towards the test focus on chemistry and science without human and physical involvement. That in turn would be aimed at a positive vision of a chemistry discourse interested in a close perspective and use of the physical and embodied human.

In many positions, issues of armed conflicts involving oil production, atomic bombs and nitroglycerine highlight devastating effects. The critical part of the figuration is directed towards the omission in the chemistry discourse of the negative consequences of chemistry discoveries/inventions and its lack of links to humans. It provides a positive vision of teaching from ethical and interdisciplinary perspectives, with connections to other cultures and global activities that not only affect ourselves and our world but also the whole world. This possible and positive vision also aims to achieve a cyborg perspective. We are all cyborgs, machine humans in which the human part must take ethical and moral responsibility to prevent the machine from taking over.

Environmental issues are found, in almost all positions, but discussed in different ways across the positions: anthropocentric most of the time and sometimes tinglified. That is, people should not be criticised for environmental problems but oil should. Sometimes, however, the blame is on us humans. Regardless of who receives the blame or whether the relationship to nature is anthropocentric or biocentric, the interpretation of the criticism in this figuration is directed towards the weak position of the environmental issues in the test and the absence of nature. I see a positive vision of this figuration that environmental issues should be studied from a broader perspective, a sustainable society where environmental issues are also linked to economic and social aspects, as many student texts in *the critical positions* do. That is, it is important to discuss and shed light on complex but difficult questions, such as one that many of the student texts have expressed: what happens when oil runs out, for me, for nature and for society?

The human body is central in most texts in the *scientific positive and critical* as well as *critical position*. Sometimes, this becomes pronounced by the writer when she/he does not answer the first three sub-items (a, b, c). The body takes, in other words, a lot of space, both in the literal and figurative sense, as it usually does in a 15-year-old's life. The body's prominent role in the texts speaks against the disembodied intellect that the tests appear to favour in the adoption of Cartesian dichotomies. The criticism of the figuration also focuses on the tests' absence of chemistry in relation to the body. The positive vision is then that teaching and learning should be embodied (Brickhouse 2001; Milne and Rubin 2011). The body is a prerequisite for our ability to analyse and understand (Alsop 2011). Moreover, chemistry in the body or in body-related chemistry (medicine, drugs, alcohol, hygiene products, etc.) is an everyday topic that is found in many student texts. Therefore, it should be an important focus of chemistry teaching. In other words, it involves ensuring that "Learning is not just about what the mind can do on its own; it is how the body, and spirit interact to create a movement, the idea of embodied learning" (Milne and Rubin 2011, p. 630).

Finally, the subject positions that appear to express resentment towards natural science and technology, those that are strongly criticising or alleged and those silently protesting in the scientific critical position and the reserved positive position depict a figuration that criticises the chemistry discourse elitist attitude of exclusion. Only the "right one" may enter this epistemological community. This then leads to a positive vision of a chemistry discourse that embraces all students and their visions, not only those with obvious scientific dreams.

13.9 Some Final Thoughts

These above-described figurations speak to us, especially to those who have influence on science teaching, including national examinations, textbooks, curriculum and syllabus. If we want students to be interested in science, we must listen to them and highlight their interests and commitments that obviously exist. We need to develop a chemistry teaching that is opposed to the one built on Cartesian thought, as the one shown in the Swedish national test in chemistry. In other words, it involves a feminist teaching and learning that is "... empowering, ... close to the women's everyday life experiences and if it builds upon the intellectual, emotional and cultural resources the participants bring to their social space" (Stromquist 1993, p. 7). This quote is from a talk that Nelly Stromquist gave at a UNESCO seminar on women's education. However, this should apply in all teaching contexts, regardless of gender, ethnicity or class. Feminist teaching and learning are also about embodiment, unlike the science inspired by Cartesian thought that separates the mind from the body. Instead, feminist-inspired science teaching can entail "... arguing for the view from a body, always a complex, contradictory, structuring and structured body, versus the view from above, from nowhere from simplicity" (Haraway 1988, p. 589).

References

Ah-King, M. (2013). Queering animal sexual behaviour in biology textbooks. *Confero, 1*(2), 46–89. doi:10.3384/confero.2001-4562.13v1i21d.

Alaimo, S., & Hekman, S. (Eds.). (2008). *Material feminisms*. Bloomington/Indianapolis: Indiana University Press.

Alsop, S. (2011). The body bites back! *Cultural Studies of Science Education, 6*, 611–623. doi:10.1007/s11422-011-9328-4.

Arvola Orlander, A., & Wickman, P.-O. (2011). Bodily experiences in secondary school biology. *Cultural Studies of Science Education, 6*, 569–594. doi:10.1007/s11422-010-9191-4.

Barad, K. (2007). *Meeting the universe halfway. Quantum physics and the entanglement of matter and meaning*. Durham/London: Duke University Press.

Berge, B.-M., & Widding, G. (2006). En granskning av hur kön framställs i ett urval av läroböcker. [An examination of how sex is produced in a section of textbooks]. Underlagsrapport till Skolverkets rapport "*I enlighet med skolans värdegrund?*". Umeå universitet.
Braidotti, R. (2002). *Metamorphoses. Towards a materialist theory of becoming*. Cambridge: Polity Press.
Braidotti, R. (2011). *Nomadic subjects. Embodiment and sexual difference in contemporary feminist theory*. New York: Columbia University Press.
Brickhouse, N. (2001). Embodying science: A feminist perspective on learning. *Journal of Research in Science Teaching, 38*(3), 282–295. doi:10.1002/1098-2736(200103)38:3<282::AID-TEA1006>3.0.CO;2-0.
Butler, J. (2007). *Genustrubbel: feminism och identitetens subversion* [Gender trouble: Feminism and the subversion of identity] (Suzanne Almquist, Trans.). Göteborg: Bokförlaget Daidalos AB. (Original work published 1990).
Butler, J. (2011). *Bodies that matter. On the discourse limits of "sex"* (2nd ed.). London/New York: Routledge.
Englund, T. (1997). Undervisning som meningserbjudanden. [Teaching as an offer of meaning]. In M. Uljens (Ed.), *Didaktik: teori, reflektion och praktik* (pp. 120–145). Lund: Studentlitteratur.
Folkeryd, J. (2006). *Writing with an attitude. Appraisal and student texts in the school subject of Sweden*. Doctoral dissertation, Uppsala University, Department of Linguistics and Philology.
Foucault, M. (2008). *Diskursernas kamp/Michel Foucault: texter i urval av* Thomas Götselius och Ulf Olsson [The discourse of struggle.] (Lars Bjurman and Jan Stolpe Trans.) Stockholm: Brutus Östlings Bokförlag.
Fox Keller, E. (1977). The anomaly of woman in physics. In S. Ruddick & P. Daniels (Eds.), *Working it out* (pp. 78–91). New York: Pantheon Books.
Gee, J. P. (2014). *An introduction to discourse analyses: Theory and method* (4th ed.). London/New York: Routledge.
Halliday, M. A. K., & Martin, J. R. (1993). *Writing science. Literacy and discursive power*. London: Falmer Press.
Haraway, D. (1988). Situated knowledges: The science question in feminism and the privilege of partial perspective. *Feminist Studies, 14*, 575–599.
Haraway, D. (1989). *Primate visions. Gender, race, and nature in the world of modern science*. New York/London: Routledge.
Haraway, D. (1991). *Simians, cyborgs and women. The reinvention of nature*. New York: Routledge.
Haraway, D. (1992). The promises of monsters: A regenerative politics for inappropriate/d others. In L. Grossberg, C. Nelson, & P. Treichler (Eds.), *Cultural studies* (pp. 295–337). New York: Routledge.
Haraway, D. (1997). *Modest_Witness@Second_Millenium. FemaleMan©_Meets_OncoMouse™*. New York/London: Routledge.
Harding, S. (1986). *The science question in feminism*. Ithaca/London: Cornell University Press.
Hasse, C. (2002). Gender diversity in play with physics: The problem of premises for participation in activities. *Mind, Culture, and Activity, 9*(4), 250–269. doi:10.1207/S15327884MCA0904_02.
Klafki, W. (1997). Kritisk, konstruktiv didaktik. [Critical constructive theory]. In M. Uljens (Ed.), *Didaktik: teori, reflektion och praktik* (pp. 215–228). Lund: Studentlitteratur.
Knorr Cetina, K. (1999). *Epistemic cultures: How the sciences make knowledge*. Cambridge, MA: Harvard University Press.
Latour, B. (1999). *Pandora's hope. Essays on the reality of science studies*. Cambridge: Harvard University Press.
Lemke, J. L. (1990). *Talking science: Language, learning, and values*. Norwood: Ablex, cop.
Lemke, J. L. (2011). The secret identity of science education: Masculine and politically conservative? *Cultural Studies of Science Education, 6*(2), 287–292. doi:10.1007/s11422-011-9326-6.
Lykke, N. (2009). *Genusforskning: En guide till feministisk teori, metodologi och skrift*. [Gender studies: A guide to feminist theory, methodology and texts]. Stockholm: Liber.

Miller, P. (2006). Contemporary perspectives from human development: Implications for feminist scholarship. *Signs, 31*(2), 445–469. doi:10.1086/491680.
Milne, C., & Rubin, K. (2011). Embodying emotions: Making transactions explicit in science education. *Cultural Studies of Science Education, 6*, 625–633. doi:10.1007/s11422-011-9354-2.
Östman, L. (1995). *Socialisation och mening: NO-utbildning som politiskt och miljömoraliskt problem*. [Socialization and meaning. Science education as a political and environmental-ethical problem]. Doctoral dissertation, Uppsala University: Department of education.
Östman, L. (1998). How companion meanings are expressed by science education discourse. In D. Roberts & L. Östman (Eds.), *Problems of meaning in science curriculum* (pp. 54–70). New York: Teachers College Press.
Pickering, A. (1995). *The mangle of practice: Time, agency, and science*. Chicago: University of Chicago Press.
Roberts, D. (1998). Analyzing school science courses: The concept of companion meaning. In D. Roberts & L. Östman (Eds.), *Problems of meaning in science curriculum* (pp. 5–12). New York: Teachers College Press.
Roberts, D., & Östman, L. (Eds.). (1998). *Problems of meaning in science curriculum*. New York: Teachers College Press.
Sjøberg, S., & Schreiner, C. (2010). *The ROSE project: An overview and key findings*. Oslo: University of Oslo.
Skolverket. (2014, January 29). Nationellt ämnesprov i kemi våren 2009 och tillhörande bedömningsanvisningar. Umeå universitet: Institutionen för tillämpad utbildningsvetenskap. Retrieved from http://www.edusci.umu.se/np/nap/tidigare-givna-prov-lpo94/
SOU. (2004:103). *Lärande för Hållbar utveckling*. [Learning for sustainable development]. Stockholm: Fritzes offentliga publikationer.
Stromquist, N. (1993). The practical and theoretical bases for empowerment. Paper presented at the international seminar on Women's education and empowerment, UNESCO Institute for Education, Hamburg, Germany.
Ståhl, M., & Hussenius, A. (2016). Chemistry inside an epistemological community box! – Discursive exclusions and inclusions in Swedish National tests in Chemistry. *Culture Studies of Science Education, 1*(11), 1–29. doi:10.1007/s11422-016-9730-z.
Ståhl, M., & Folkeryd, J. (2016). Scientific norms and students' evaluative language use. In M. Ståhl, *Kemiämnets normer och värden. Diskursanalytiska studier av nationella prov i kemi och tillhörande elevtexter*. Doctoral dissertation, Uppsala University, Department of Education.
Søndergaard, D. M. (1996). *Tegnet på kroppen. Køn: Koder og konstruktioner blandt unge voksne i akademia*. [Signs on the body. Sex: Codes and constructions among adults in academia]. Köpenhamn: Museum Tusculanums forlag, University of Copenhagen.
Wenger, E. (1998). *Communities of practice*. New York: Cambridge University Press.
von Wright, M. (1999). *Genus och text - när kan man tala om jämställdhet i fysiklärome-del?* [Gender and texts – When is it possible to talk about equality in physics' textbooks?] Stockholm: Skolverket.

Chapter 14
Sensory Science Education

Kathrin Otrel-Cass

14.1 Introduction

This chapter intends to discuss the teaching of scientific ways of thinking, talking and doing things at primary school that involves an explicit focus on the senses. The intention is to trouble science education pedagogy that overlooks that young people/we all experience the world through our body *and* our mind. Such pedagogy may need to conceptualize how people make sense of the world. Although the body with its senses is our main medium to perceive and act on the world, sensorial experiences have been described as distractions, and this has been critiqued by philosophers like Shusterman (2012). With this in mind, this chapter will focus on the idea and place for a sensory science pedagogy. Bronwen Cowie, Judy Moreland and Kathrin Otrel-Cass (2013) described that science learning outcomes can be unpacked into conceptual, technical, procedural and societal learning domains, yet conventional science education pedagogies continue to propose learning activities that focus on the mind. When I say mind I mean that there is very little literature that refers explicitly to the possibilities for embodied and sensorial learning and how such approaches could contribute to science education. For instance, while the Eurydice report (2011) on science education in Europe refers to the human body, it only writes about the body as a context example for learning without any further explanation what this may mean or whether there may be fertile learning opportunities that can be achieved by learning about the body through the body. My argument is further that research on science education should explore the place for sensory-based experiences, especially if and how individuals can make meaningful connections between their sensory experiences and the world of scientific explanations.

K. Otrel-Cass (✉)
Department of Learning and Philosophy, Aalborg University, Aalborg, Denmark
e-mail: cass@learning.aau.dk

K. Otrel-Cass et al. (eds.), *Cultural, Social, and Political Perspectives in Science Education*, Cultural Studies of Science Education 15,
DOI 10.1007/978-3-319-61191-4_14

I will discuss the concept of sensory pedagogy and its relevance to learning experiences by presenting empirical examples that originated from ethnographic work conducted in different primary classrooms. This chapter will elaborate on how a sensory approach in science education research may contribute to the study of primary classroom science education. This chapter emphasizes how a theorized focus on sensory pedagogy is still underexplored in science education and that this involves considering the different kinds of materials that may be encountered therein. I will focus particularly on the process of these experiences and the connection with the material ecology that shape these processes. This chapter's focus is on the utilization of a sensory approach to science education since it is central to our understanding of the world. The argument made here is that a sensory pedagogy is still underexposed. Science education has come a long way, for example, Wolff-Michael Roth (2007) looked at how the individual can make meaningful connections with the world and develops an identity in science, while Kenneth Tobin (2012) discussed that learning science should be a collective activity. Investigations about embodied science learning include studies that explain how to collect evidence by taste or smell, for example, the work by Richard Blatchly, Zeynep Delen and Patricia O'Hara (2014), while William Stansfield (2012) explains how deceptive the senses can be and that information derived through the senses can be problematic since they may produce conflicting information. Investigating the role sensory experiences play is an acknowledgement that our senses are central to our perceptions of the world and that research needs to pay attention to understanding sensoriality. These arguments by Sarah Pink (2015) have also been picked up by Eva Johansson and Gunvor Løkken (2014) who argue for the adoption of a sensory pedagogy in pre-schools. They underpin their argument on the ideas of the French philosopher of the human body Maurice Merleau-Ponty (1968). Merleau-Ponty's philosophical angle is a reconceptualization of the Cartesian argument which is focused on the mind, to explain how the materiality of the world is experienced through the body and the mind. Continuing these discussions into the specifics of science education is the basis of this chapter.

Before going into more detail of sensory pedagogy in science education, I will start by positioning myself and my current views on learning. I am departing from within a sociocultural perspective to science learning. This implies that motivation, expression and development can be understood as being mediated by social, material, temporal, historical and cultural aspects as was argued by Lev Vygotsky (1981) or James Wertsch (1991) and that learning cannot be disassociated from the context within which it takes place (Cowie et al. 2013). Learning in (primary) science education implies that young people also develop an appreciation of the culture of science to understand how scientists see and interpret the world and generate knowledge. This has been much discussed and is typically described as learning about the nature of science (NOS). William McComas and Hiya Almazroa (1998) also emphasized that scientific knowledge creation has many tentative and exploratory facets. However, the conceptualizing of perception and learning in the world through the body and the mind requires also a consideration of culture (Shusterman 2012). Culture, as Shusterman explains, provides 'language, values, social institutions and artistic media through which we think and act' (p. 27). But culture also shapes what we consume or how we work with our bodies. Shusterman uses also the term body-mind developed from John Dewey's conceptualizations. I will use the term body-mind in this chapter as it nicely unifies – rather than separates – the two. Another useful metaphor of body-mind is that of a distributed network that is in an ongoing exchange with the external world. Elisabeth Osgood-Campbell (2015) explains that this includes dealing with the basic mechanisms of moving around to the intricate ways of understanding the world. A consequence of these ideas is that the body-mind togetherness plays a primary role for learning when it comes to identifying the challenges that this unit presents.

Investigating the sensory dimensions to experiencing the world and how to link this to science education and its materials in the primary classroom is based on the ideas of philosopher Maurice Merleau-Ponty (1962, 1968) who pointed to the significance of embodied experiences with the world and its constituents and its impact on engagement with the world. Exploring sensory-based pedagogies requires to understand the opportunities but also challenges when students experience and make connections through their bodies and with their senses. Of course, it has been long argued that the body-mind is integral to learning (Dewey 1963), while Tracy Thomson writes that 'learning does not happen exclusively above the neck' (2015, p. 6). However, a body-mind pedagogy for science education needs to address this epistemology (dualism of thought) and experience (classroom practice), especially when the two are at times at odds. Scott Hamilton and Trevor Hamilton (2015) provide the example when the mind is being deceived into producing scientifically incorrect interpretations about what is happening to its body. Body-mind pedagogy in science will need to address the interconnections and confusions that occur about the mind's image of the body in a material space.

To illustrate and ground the argument for a body-mind pedagogy in science education, examples will be presented from primary science classes from several studies that involved young people between the age of 8 and 12 years in New Zealand and Denmark. These examples intend to illustrate theoretical concepts and to make the reader reflect on their own body-mind experiences.

The rest of this chapter is organized in the following way: the continued discussion will briefly cover sensing, senses and the *sensorium* (a term that describes the entire apparatus of senses). It continues with a focus on why, in science education, it is important to pay attention to the senses and the challenges of a body-mind pedagogy when the world is to be understood in scientific ways. This chapter is based on making sense of children's experiences that were captured on video during classroom observations. It will also address the implications for sensory research methodologies. A discussion on body-mind pedagogy in everyday science classrooms will round off this chapter.

14.2 The Sensory Trap: Seeing, Hearing, Touching and Tasting Science

Vignette:

> A long time ago now I used to conduct educational workshops for local salt mines. Historically the local rock salt used to be extracted from the mountains by being first dissolved in water and then pumped down the mountain to be later cooked in big pans until all water had evaporated leaving the rock salt behind. In the workshops I repeated this process and dissolved salt in water before offering young visitors to dip their fingers into the salty water before letting the water evaporate in a pan. The response from children was always one of delight – the water tasted salty although the salt had just seemingly disappeared in front of their eyes. Tasting the water could make the children 'see' the salt.

The five senses, sight, hearing, touch, smell and taste, typically describe our modes of experiencing the world. Despite the familiarity we all have with our senses, their relevance to learning is still under-researched. As Don Ihde (2007) notes, nothing seems easier than reflecting on your own experiences because of the familiarity with our own experiences, but on the same token, it is exactly this familiarity that makes it very difficult for us to reflect on them. When it comes to thinking about the senses, it is useful to consider Aristoteles (1906) common sense descriptions. He highlights in his work *De Sensu and De Memoria* that senses are oriented towards the objects they are engaged in. He explains that sight, as an example, is shaped by the object of the thing that is been looked at and mediated by the rays between the eye and the object. The Aristotelian argument is therefore that a sense needs to be conceptualized with the context of the object it is oriented towards. While this description has been criticized for being not defined enough, see, for example, Richard Sorabji (1971). It may be a useful way to see sensing as a material's oriented account of perception. However, we don't perceive things in isolation but rather operate a sensorium or apparatus of sensing to 'mediate the relationship between self and society, mind and body, idea and object' as Michael Bull, Paul Gilroy, David Howes and Douglas Kahn point out (2006, p. 5).

Teresa de la Isla (2008) explains further that in addition to the five senses that are commonly known, there are two additional senses: the proprioceptive sense, the sense of location in space, and the vestibular sense, the sense that detects motion. I will however concentrate for now on the five senses and their relevance to science education.

The First Sense: Sight Seeing and observing things in science are important skills and may also include that 'invisible' aspects are made visible for example through models or other representations. In science, this requires that young people develop 'metavisual capability' as John Gilbert (2004) explains. Seeing things in science is more than 'everyday seeing', because what scientists see needs to be converted into a scientifically acceptable understanding about the world:

> Science seeks to provide explanations for natural phenomena: to describe the causes that lead to the particular effects in which scientists are interested. However, phenomena are not ready – made: we impose our ideas of what might be important on the complexity of the

> natural world. Scientists then investigate these idealisations, what may be called 'exemplar phenomena', at least at the outset of the enquiry is in any given field. (Gilbert 2004, p. 10).

Seeing things in science is often achieved by extending what is possible through technological tools, using tools including microscopes, telescopes or X-ray to go beyond what we can see under 'everyday' conditions to expand what our visual senses are equipped to handle. This means when thinking about the sense 'sight' and its relevance to a body-mind pedagogy, it is important to distinguish between seeing and making things visible, even or especially those things that are impossible to see following direct observation. Science education is overlaying seeing and observing with rational judgement, and this may mean not to always trust our eyes and to retrain what we can make from seeing and how to interpret this. For example, we all learn at some stage that the world does not end where our eyes can make out the horizon, the line that seemingly separates the Earth from the sky. We learn to distrust our eyes to complement our understanding with the information we gather from secondary information in imagery (e.g. pictures of the Earth taken from space) or information that is shared verbally or in text by others.

The Second Sense: Hearing Hearing, sound and aurality have as much to do with the nature of sounds and soundscape (the environments and sources that produce sound) as well as the culture of listening as clarified by Catherine Burke and Ian Grosvenor (2011). When it comes to sound at school or more specifically in science education, hearing is often about making auditory observations, such as listening to the sound of birds or understanding the difference between volume and pitch.

Imagine teaching the Doppler effect without providing an opportunity for students to listen to the changing pitch of a moving sound-producing object. Teachers regularly utilize the hearing sense; however, it is not necessarily a common practice to go beyond listening to explore, for example, the complexity of acoustic ecology. Paulo Simeão Carvalho, Edite Briosa, Marcelo Rodrigues, Carla Pereira and Margarida Ataíde (2013) argue further that hearing activities have no great prominence in modern science education, perhaps also because the science of sound utilizes imagery such as waves to explain what it is and there is more often a focus on how to replace aurality of sound with the visuality of sound.

The Third Sense: Touch Touch is perhaps at first mostly associated with the use of our fingers. Equipped with pressure and temperature sensors, they provide feedback to our brain about the things we can feel. Touch is also about how we respond to and experience the world through our feet; see, for instance, the work by Alen Hajnal, Daniel Abdul-Malak and Frank Durgin (2011). Interest in haptic sensorial dimensions has increased with the rise of computer simulations and games highlighting how significant touch is when we are perceiving the world. Tim Ingold (2000) describes that through touch, our bodies experience the surfaces of the material world around us. Feeling the ruggedness of a rock provides us with sensorial information on what constitutes stoniness.

The Fourth Sense: Smell Utilizing the sense of smell can support, for example, the teaching of chemistry being the science of the structure, composition and properties of matter. Deborah Bromfield-Lee and Maria Oliver-Hoyo (2007) share their insights when they found that utilizing smell could function as an analytical tool since the activity involved working with esters that have naturally occurring detectable aromas. Bruno Latour (2004) wrote about 'becoming a "nose" [un nez]' (p. 77) and that working with your senses such as smell requires that we need to learn how to be affected by the stimulants our body gets in contact with also to distinguish difference and become more self-aware of the body's abilities.

The Fifth Sense: Taste The language of science supplies us with pictures to be seen when nothing or perhaps even contradictory things can be seen. In my earlier example, I referred to preparing a salt solution. The water tasted salty, yet it still looked clear. Only when seeing was complemented through taste – the gustatory sense – made things visible that were not apparent previously with the eyes only. Taste can therefore be utilized to identify chemical signatures of things to provide information that is complementary yet different to what we can see, hear or feel.

The Sensorium: Working with the Senses The sense-making or understanding we create of the world is created by being in the world through the entirety of the body-mind. The different senses complete each other to provide richer and more meaningful insights. Seeing with our ears include the pictures we produce, for example, when a car drives by that we haven't observed necessarily/directly or seeing a bird when we hear a bird's song. Seeing through touch allows us to translate patterns into pictures. This is also called haptic visualization where micro-details contribute to fill the details of a visual whole. Miriam Reiner (2008) explains that touch generates images in our brain of shapes and materials, for example, if something is metallic or made from wood, and this in turn triggers ways of operating and responding to those materials. The process from seeing and observing to involving the other senses to see creates each time complex new layers to what we understand about the world. This can result also in significant differences as to how things are perceived and/or handled as James Brockmole, Christopher Davoli, Richard Abrams and Jessica Witt (2013) clarify: 'Whether one takes an object in the hands or manipulates an object with a tool, profound changes in perception, attention, and memory are observed' (p. 38). This potential complementarity of the senses to enhance body-mind experiences has been argued for to be utilized for science teaching by Deborah Bromfield-Lee and Maria Oliver-Hoyo (2007).

The consequences of how we experience and perceive the world through our senses are extraordinary and mundane at the same time. It means that without supplementing our visual sense, we would process information inadequately, but it may also be the reason why it can be problematic to overcome perceptual information and to replace it with 'scientifically correct' explanations (see, e.g. studies conducted on the topic of misconceptions in science by Rosalind Driver in 1989 and many more). The basis of all this is that we form experiences based on the multisensorial encounters with the world. Since the world around us is an assemblage of different things including natural and man-made, encounters with the materialized world form part of the continuous experiences we have. However, how can those insights inform the practices that play out in science classrooms?

14.3 Pedagogical Approaches to Experiencing the Materialized World

Learning is an active process that results in some kind of transformation. This could be the acquisition of some new ideas connected to some exiting ones, it could encompass transformation of practices such as gaining procedural knowledge that identifies sequences of an operation, and it could involve acquiring some technical skills, for example, being able to carefully measure and extract particular amounts of liquid in an experiment. Learning is also a function from being together with people and using tools and using Vygotsky's words, 'learning is a necessary and universal aspect of the process of developing culturally organized, specifically human psychological function' (1978, p. 90). By involving or exposing oneself in and with the world, experiences are made. Insights and knowledge result from this exposure and may allow for deductions about the world that are based on those experiences. Following the arguments by Vygotsky, Dewey and more recently Wolff-Michael Roth and Alfredo Jornet (2014), experiences are entangled with emotions or affect, so the encounters we have mean that we are 'subject and subjected to experience' (p. 3) and 'qualities that shape the experience, transform it, and are, in turn, transformed by the unfolding experience' (p. 10). Therefore, the sensorial experiences with the materialized world produce not only representations in our mind, but they also produce feelings and emotions. Following this argument, Roth and Jornet propose a theory of experience to say that experience manifests itself in/as passions (affect, emotion) and that experiences are integrated in the explanations and understandings we have about the world over space and time. Experience, they say, is thus 'a moving force' and results in personalized transformation.

A pedagogical framework that aims to acknowledge and address that we perceive and make sense of the world in different ways is that proposed by Howard Gardner (1999). He advocates catering for eight different intelligences and utilizing different learning activities that allow for visual, kinesthetic, creative and sociocultural encounters that promote interactions and engage learners in learning experiences. Howard Gardner and Thomas Hatch explain that teachers need to cater for different kinds of intelligences and 'the capacity to solve problems or to fashion products that are valued in one or more cultural setting' (1989, p. 5). There has also been criticism about this view that draws perhaps all too simplistic conclusions such as that some people are simply visual learners while others may be, for instance, verbal learners. This indicates that there is no rigidity in how we learn and that we can improve our ability to use our senses through training, as explained by Nora Newcombe and Mike Stieff (2012).

Richard Felder and Linda Silverman (Felder 1993; Felder and Silverman 1988) point at four dimensions to be considered for a pedagogy of the senses, including (1) the sensory or intuitive, the preferred way of receiving information; (2) the visual or verbal, the mode on how to receive information; (3) actively or reflectively, describes the process of how information is being received; and (4) sequentially or globally, refers to the sequence in receiving information. While this pedagogical framework is useful in addressing diverse ways of learning, its limitations lie in its 'either or' approach. The danger is also, as Ingold (2000, p. 284) points out, that we reduce 'the body to a locus of objectified and enumerable sense whose one and only role is to carry the semantic load projected onto them by a collective, supersensory subject – namely society'. Ingold (2000) suggests instead to focus on 'the creative interweaving of experience', and as Pink argues 'no sensory modality necessarily dominates how…practices are experienced' (2015, p. 11).

Taken from this, the argument can be made that a body-mind pedagogy in science education needs to offer a broader approach to accommodate for the interwoven sensorial experiences we make. Learning takes places not only through the mind but through the body-mind, with its skilled subjectivity that helps our everyday knowledge construction (Shusterman 2012).

Body-mind pedagogy is about the intricate relationship we have with materials. We express and identify characteristics of the physical world around us by values that we put on descriptions of materials after we have had a sensorial experience with them. We like or dislike the 'feel' of something. Elvin Karana, Paul Hekkert and Prabhu Kandachar (2009) explain how people attribute expressive meanings such as materials being robust. They describe further that experiences with materials are attached to emotions – food (materials) may be a useful example. We attribute sensorial properties to materials. Something can be described as being smooth, shiny or cold (Karana et al. 2009). Situated in the context of manufacturing processes, the authors argue that the meaning we attribute to materials can vary – it is not only the material itself but also the object it embodies – how we encounter or perhaps use the material and any prior experiences we may have had with similar materials. This suggests that a body-mind pedagogy should consider materials and their materiality, meaning the context and conditions on how materials are experienced.

Next, I will argue that researching sensorial experiences to promote an argument for a body-mind pedagogy requires also a careful examination of the methodologies for such research.

14.4 Utilizing Sensory Methodologies

Researching sensory-based experiences also requires a sensitized approach to the way data is being collected and analysed. Following Gillian Rose's introduction to the interpretation of visual materials (2007), I adopt a critical approach to the interpretation of body-mind experiences in the material world of the science classroom. By that I mean to be careful about my own interpretations of visualizations when I have captured experiences through the senses on video. Rose reminds us that modernity and postmodernity are ocularcentric because we interact more with the visually constructed experiences. A methodology on visuality and the senses requires seeing the world through more than just the eyes, to go beyond the production of visual hierarchies that create order and differences, as clarified by Donna Haraway (1991), a methodology that adopts a critical perspective 'differentiates between the social effects of those different visions' (Rose 2007, p. 5).

I want to draw attention on the modalities of how the materialized world is experienced in science education classrooms and how it can be interpreted. A focus needs to be put on the interpretation of what is being witnessed since educational researchers reinterpret. To understand what this kind of perspective incorporates, it may be easier to explain what it's not. For this argument, sensory experiences that are visually captured are more than just the background context of goals, intentions and activities in which an embodied performance has been captured in a Ludwig Wittgenstein's (1967) sense. I also do not wish to build further on Pierre Bourdieu's (1984) notion of habitus that describes the body as background to an individual's disposition, perception and behaviour through which one perceives the world. I am interested in identifying the instances where body-mind sensations can temporarily disorient perspectives and perhaps disturb the engagement with the world. These instances may be easier to locate and unpack. This perspective is also argued for by philosopher Maurice Merleau-Ponty (1968) whose argument is that one cannot simply observe one's own body or reflectively observe oneself, because one can only observe external objects through one's body.

The sensorial encounters that I witnessed in classroom observations on a number of occasions are of course reinterpretations and represent my own experiences with those who had the experiences in the first place. However, the examples that will be presented should not necessarily represent a first-person account of sensorial body-mind experiences but rather provide different scenarios to continue and build the discussion about a body-mind pedagogy in science education.

14.5 Examples of Sensory Classroom Experiences

In this section, I am presenting examples of sensory experiences in the science classroom. Each example starts with a brief explanation of the origin of the empirical data. All of the projects involved the collection of video recordings of classroom activities.

Example 1 – Research project: Linking culturally responsive teaching, learning and assessment to enhance the engagement of diverse students in primary science classrooms. Summary report: Bronwen Cowie, Kathrin Otrel-Cass, Ted Glynn and Helena Kara, with Marion Anderson, Judy Doyle, Asri Parkinson and Christina Te Kiri (2011). Data collection included classroom observation using video, photographs and field notes as well as interviews, conducted over a 2-year period.

14.5.1 *Feeling the Temperature of the Water*

In her unit about weather, Julia wanted the children to learn about the concept of temperature. She asked her students to feel the water in two bowls, one filled with cold water the other one with warm water. Julia invited the year 4 (9-year-old) children in her New Zealand classroom to test the water and describe their experiences. This activity was met with much interest and enjoyment. Many children wanted to volunteer to try out feeling the water, to see whether they agreed or not with the previous child on how they experienced the temperature with their hands. The children did not necessarily agree on their experiences, but this was not the point but rather to take note of how warm or cold the water feels (Fig. 14.1).

Following up from the experiments with the children, Julia wrote on an A3 sheet (Fig. 14.2).

The teacher integrated the descriptions used when the children explored the water temperature by using their hands. She later compared it to using a thermometer, also to link between what feels warm or cold and what the thermometer shows. The meaning that was attributed to the temperature of the water focused on what could be felt. Water was not just water but warm, cold or really cold water, thus was attributed through temperature sensing of the children's hands. Not all children agreed either drawing attention to the subjective and unique experience of the individual.

14.5.2 *Augmented Reality – Observing the Sun and the Moon: Virtuality Versus Reality*

Example 2 – Research project: 'Networked Inquiry Learning in Secondary Science' classroom study (NILSS) was originally conducted in New Zealand (2010–2012) and also a Danish school is included in 2012. Data collection included classroom observation using video, photographs and field notes as well as interviews. The Danish case presented here was conducted over a 6-month period.

Poul, a Danish grade 9 (14-year-old) teacher, supported his class who were working on an inquiry-based activity. The children were investigating their own questions about space and the universe. When Laura and Line showed him an app on their phone that allowed them to see the position of the sun and the moon in space, Poul gave the two girls welder's glass to observe the sun directly outside the classroom. The girls were amazed at their observation (Figs. 14.3 and 14.4).

Seeing with their own eyes through the welder's glass attributed a sense of reality to the existence of the sun, even though the sun is there all the time and even though they had 'seen' the sun through the app on their phone. The experience made possible through the glass attributed awe and wonder to the experience of seeing the sun. It added drama to the girls' immediate world, and in a sense, they brought the sun to life (Ingold 2000).

Fig. 14.1 Feeling the water

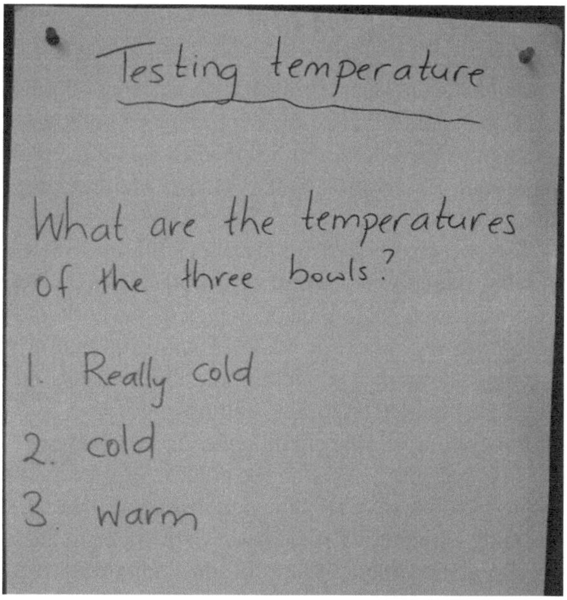

Fig. 14.2 Temperature chart

Fig. 14.3 Looking at the sun and moon on a smartphone

14.5.3 Touching the Cork: Experiencing Materiality and Why Frustration Can Be Good

Example 3 – Research project: 'Ny naturfag – Fremtidens naturfagslokale' (2014–2015) is conducted at a Danish primary school to investigate how the physical environment shaped the teaching of science and technology. Data collection included classroom observation using video, photographs and field notes as well as interviews.

The two girls were working with a plastic wine cork, hot glue and plastic wings to build the spinning wheel of a water wheel. It took the girls nearly an hour of frustrated attempts to glue the wings on the cork until they realized that it wouldn't work with this material but only with the natural cork. Their attempts included touching the cork and sticking the three wheels on carefully only to realize that when the hot glue dried, they fell off.

Frustration was the emotion the girls encountered when they were working with the material plastic cork in the context of constructing a water wheel. The plastic versus the cork were slight material qualities that the girls only detected after experimenting with them. Perhaps they were drawn to the smooth look of the plastic cork, only to realize that attributes of glueability they needed for this activity were only true for the cork.

Fig. 14.4 Seeing the sun

14.6 Putting Forward Sensory Pedagogies for Science Education

To utilize a sensory pedagogy, it will be important to consider the unity between the body and the mind, but for thoughtful planning, this requires a recognition that while the two parts are entangled, they are not always harmonized (Shusterman 2008). Cultural and social conditions impact on how the body-mind experiences are configured (Dewey 1989). In the following, I am synthesizing three key ideas.

A body-mind pedagogy that allows for individual, subjective and emotional experiences because bodymind encounters with the material world are had by individuals and produce emotions (Roth and Jornet 2014). When the children were trying out and testing the temperature of the water, they were excited laughing and describing what they felt, agreeing and disagreeing with each other, and this was based on making the same and yet not necessarily having the same experiences. Laura Malinverni and Narcis Pares (2014) explain that embodied learning necessitates a pedagogy that emphasizes learning by doing but that it also requires an understanding that the body plays a central role in the processing of mental models.

A body-mind pedagogy in science that considers materials and material ecology. Materials are experienced in environments that are populated through objects (Ingold 2000). Even subtle changes in the physical configuration of the environment can have relevant impact on comprehension pointing out the need of fine-grained analysis of the relation between design choices behaviour and cognitive processes (Malinverni and Pares 2014). The girls had chosen to work with the plastic cork on purpose. It was through their hands-on work with the materials that allowed them to explore the characteristics of the plastic cork as a material for building a water wheel. Once they tried the natural cork, it became clear to them that the material properties were superior to achieve their goal. The body-mind performances and interactions with the material world produce also an effect on others, as written by SungWon Hwang and Wolff-Michael Roth (2007); in this example, the other children who were watching Laura and Lena suggested to the girls to try a different cork.

Realizing the different modes available for meaning-making in a body-mind pedagogy. Carey Jewitt (2009) explains that it is necessary to consider the characteristics of different modes to understand how meaning is made and communicated. The body-mind experiences of the girls looking at the position of the sun using their phone and later the welder's glass produced and affected their understanding. The teacher's own body-mind memory was prompted when he saw the sun's representation on the phone, and he identified an opportunity for the students to connect one mode of representation with another mode of experiencing an observation of the sun in a safe way. It is the coming together of emotional and sensorial encounters that create complex 'pictures' in our minds of those experiences. Paying attention to our body-mind memory in addition to paying attention to how emotions and our senses shape what is being experienced is about subjectivity of body-mind.

14.7 In Conclusion

Our mind transforms and objectifies the experiences we have with the world. Dewey explained that this takes place through the increasing levels of engagement with the world and the quality of the experiences (1963). However, experiences even when they occur for the same group of people are lived in different ways. The body-mind unit that is shaped by social conditions and material encounters is one that is also based on body-mind memory. The experiences we have are thus unique, are shaped by the past and will shape future body-mind encounters. Of course, the body-mind experiences we make are not all educational, that is, with an educational purpose in mind. But placed under educational conditions, they become experiments in a Deweyan sense where experiences are about trying out things through and with the body-mind. Historical and sociocultural conditions, the interactions with a material world, create realities for individuals that are situated in their body-mind constructions. A sensory pedagogy in science needs to address that utilizing the senses requires a deliberate sensitization and validation of the presence of the senses. They are not entities that can be turned on and off, and learning is not a prerequisite to having sensorial experiences. The sensorial experiences we have in the world are there regardless, and it is a matter of giving them space in the science classroom.

A sensory pedagogy approach needs to take note of the unique ways on how young people/we all experiment and experience the world. The material and natural world that is the subject of investigation for science is thus experienced emotionally and sensorially only to be evaluated by the body-mind based on context and prior experiences. Troubling science education pedagogy means here to reconceptualize who we are and how we make sense of the world and the recognition that the body-mind is present, imbalanced and complex.

References

Aristotle, Ross, G. R. T., & Aristotle. (1906). *Aristotle De sensu and De memoria*. Cambridge: University Press.

Blatchly, R. A., Delen, Z., & O'Hara, P. B. (2014). Making sense of olive oil: Simple experiments to connect sensory observations with the underlying chemistry. *Journal of Chemical Education, 91*(10), 1623–1630. http://doi.org/10.1021/ed300557r.

Bourdieu, P. (1984). *Distinction: A social critique of the judgement of taste*. Harvard University Press.

Brockmole, J. R., Davoli, C. C., Abrams, R. A., & Witt, J. K. (2013). The world within reach effects of hand posture and tool use on visual cognition. *Current Directions in Psychological Science, 22*(1), 38–44. http://doi.org/10.1177/0963721412465065.

Bromfield-Lee, D. C., & Oliver-Hoyo, M. T. (2007). A qualitative organic analysis that exploits the senses of smell, touch, and sound. *Journal of Chemical Education, 84*(12), 1976. http://doi.org/10.1021/ed084p1976.

Bull, M., Gilroy, P., Howes, D., & Kahn, D. (2006). Introducing sensory studies. *The Senses and Society, 1*(1), 5–7.

Burke, C., & Grosvenor, I. (2011). The hearing school: An exploration of sound and listening in the modern school. *Paedagogica Historica, 47*(3), 323–340.

Carvalho, P. S., Briosa, E., Rodrigues, M., Pereira, C., & Ataíde, M. (2013). How to use a candle to study sound waves. *The Physics Teacher, 51*(7), 398–399.

Cowie, B., Moreland, J., & Otrel-Cass, K. (2013). *Expanding notions of assessment for learning: Inside science and technology primary classrooms*. Rotterdam: Sense Publishers.

Cowie, B., Otrel-Cass, K., Glynn, T., & Kara, H., with Anderson, M., Doyle, J., Parkinson, A., & Te Kiri, C. (2011). Culturally responsive pedagogy and assessment in primary science classrooms: Whakamana tamariki. Summary Report Wellington: Teaching Learning Research Initiative.

Dewey, J. (1963). How we think. In R. M. Hutchins & M. J. Adler (Eds.), *Gateway to the great books* (Vol. 10). Chicago: Encyclopedia Britannica, Inc.
Dewey, J. (1989). Having an experience. In A. Boydston Jo (Ed.), *John Dewey: The later works, 1925–1953, Art as experience* (Vol. 10, pp. 42–63). Carbondale: Southern Illinois University Press.
de la Isla, T. (2008). Making sense out of everyday routines. *Exceptional Parent, 38*(2), 34–35.
Driver, R. (1989). Students' conceptions and the learning of science. *International Journal of Science Education, 11*(5), 481–490. http://doi.org/10.1080/0950069890110501.
Eurydice. (2011). *Science education in Europe: National Policies, practices and research*. Brussels: Eurydice.
Felder, R. M. (1993). Reaching the second tier: Learning and teaching styles in college science education. *Journal of College Science Teaching, 23*(5), 286–290.
Felder, R. M., & Silverman, L. K. (1988). Learning and teaching styles in engineering education. *Engineering Education, 78*(7), 674–681.
Gardner, H. (1999). The disciplined mind. Simon & Schuster New York. Retrieved from http://www.academia.edu/download/30301649/The_Disciplined_Mind.pdf
Gardner, H., & Hatch, T. (1989). Multiple intelligences go to school: Educational implications of the theory of multiple intelligences. *Educational Researcher, 18*(8), 4–9.
Gilbert, J. K. (2004). Models and modelling: Routes to more authentic science education. *International Journal of Science and Mathematics Education, 2*(2), 115–130.
Hajnal, A., Abdul-Malak, D. T., & Durgin, F. H. (2011). The perceptual experience of slope by foot and by finger. *Journal of Experimental Psychology: Human Perception and Performance, 37*(3), 709.
Hamilton, S., & Hamilton, T. J. (2015). Pedagogical tools to explore Cartesian mind-body dualism in the classroom: Philosophical arguments and neuroscience illusions. Frontiers in Psychology, 6. http://doi.org/10.3389/fpsyg.2015.01155.
Haraway, D. (1991). *Simians, cyborgs, and women: The reinvention of women*. London/New York: Routledge.
Hwang, S., & Roth, W.-M. (2007). From designing artifacts to learning science: A dialectical perspective. *Cultural Studies of Science Education, 1*(3), 423–450.
Ihde, D. (2007). *Listening and voice: Phenomenologies of sound*. New York: SUNY Press.
Ingold, T. (2000). *The perception of the environment: Essays on livelihood, dwelling and skill*. Milton Park/Abingdon/Oxon: Routledge.
Jewitt, C. (2009). *The Routledge handbook of multimodal analysis*. London: Routledge.
Johansson, E., & Løkken, G. (2014). Sensory pedagogy: Understanding and encountering children through the senses. *Educational Philosophy and Theory, 46*(8), 886–897. http://doi.org/10.1080/00131857.2013.783776.
Karana, E., Hekkert, P., & Kandachar, P. (2009). Meanings of materials through sensorial properties and manufacturing processes. *Materials & Design, 30*(7), 2778–2784.
Latour, B. (2004). How to talk about the body? The normative dimension of science studies. *Body and Society, 10*(2–3), 205–229. http://doi.org/10.1177/1357034X04042943.
Malinverni, L., & Pares, N. (2014). Learning of abstract concepts through full-body interaction: A systematic review. *Journal of Educational Technology and Society, 17*(4), 100–116.
McComas, W. F., & Almazroa, H. (1998). The nature of science in science education: An introduction. *Science & Education, 7*, 511–532.
Merleau-Ponty, M. (1962). Phenomenology of perception, trans. *C. Smith*. London: Routledge.
Merleau-Ponty, M. (1968). *The visible and the invisible: Followed by working notes*. Evanston: Northwestern University Press.
Newcombe, N. S., & Stieff, M. (2012). Six myths about spatial thinking. *International Journal of Science Education, 34*(6), 955–971. doi:10.1080/09500693.2011.588728.
Osgood-Campbell, E. (2015, March 1). Investigating the educational implications of embodied cognition: A model interdisciplinary inquiry in mind, brain, and education curricula. *Mind, Brain, and Education, 9*(1), 3–9. doi:10.1111/mbe.12063.

Reiner, M. (2008). Seeing through touch: The role of haptic information in visualization. In *Visualization: Theory and practice in science education* (pp. 73–84). Springer. Retrieved from http://link.springer.com/chapter/10.1007/978-1-4020-5267-5_4
Pink, S. (2015). *Doing sensory ethnography*. London: Sage.
Roth, W.-M. (2007). The ethico-moral nature of identity: Prolegomena to the development of third-generation cultural-historical activity theory. *International Journal of Educational Research, 46*(1–2), 83–93. http://doi.org/10.1016/j.ijer.2007.07.008.
Roth, W.-M., & Jornet, A. (2014). Toward a theory of experience. *Science Education, 98*(1), 106–126.
Rose, G. (2007). Researching visual materials: Towards a critical visual methodology. In *Dies., Hg., Visual methodologies. An introduction to interpreting visual materials* (pp. 1–32). London: Sage.
Shusterman, R. (2008). *Body consciousness: A philosophy of mindfulness and somaesthetics*. Cambridge, MA: Cambridge University Press.
Shusterman, R. (2012). *Thinking through the body: Essays in somaesthetics*. Cambridge, MA: Cambridge University Press.
Stansfield, W. D. (2012). Science & the senses: Perceptions & deceptions. *American Biology Teacher, 74*(3), 145–149. http://doi.org/10.1525/abt.2012.74.3.4.
Sorabji, R. (1971). Aristotle on demarcating the five senses. *The Philosophical Review, 80*(1), 55–79. http://doi.org/10.2307/2184311.
Thomson, T. (2015). Sensory-based arts education and engagement in the junior classroom: Exploring multiple ways of knowing and meaning. Doctoral dissertation, The University of Western Ontario.
Tobin, K. (2012). Sociocultural perspectives on science education. In *Second international handbook of science education* (pp. 3–17). Springer. Retrieved from http://link.springer.com/chapter/10.1007/978-1-4020-9041-7_1
Vygotsky, L. S. (1981). The instrumental method in psychology. In J. V. Wertsch (Ed.), *The concept of activity in soviet psychology* (pp. 135–143). Armonk: M.E. Sharpe.
Vygotsky, L. S. (1978). *Mind in society: The development of higher mental process*. Cambridge, MA: Harvard University Press.
Wertsch, J. V. (1991). A sociocultural approach to socially shared cognition. Retrieved from http://psycnet.apa.org/books/10096/004
Wittgenstein, L. (1967). *Zettel*. Berkeley: University of California Press.

Chapter 15
Troubling Science Education and Imagining Possibilities for Transformation: An Afterword

Christina Siry

> [O]f all our cognitive capacities, imagination is the one that permits us to give credence to alternative realities. It allows us to break with the taken for granted, to set aside familiar distinctions and definitions. (Greene, 2000, p. 3)

I start my commentary on this volume with the above quote from Maxine Greene, as I take inspiration from the notion that in and through imagination we can break with taken-for-granted assumptions. That, for me, is one of the central notions at the core of the chapters of this book – why, how, and in what specific ways can we give credence to 'alternative realities', as Greene so eloquently writes about in her book *Releasing the Imagination*, a work that has inspired much of my own thinking, researching, and writing about education through the years. Thus, in my response to the chapters herein, I use Greene's writings to underscore the potentialities that can emerge when we 'release the imagination' and consider the possibilities for alternative approaches within science education research and practice.

Writing an afterword is an opportunity, an opportunity to consider the contents of a book and to reflect on the contribution the particular book makes to the field. I am honoured to contribute to this volume, and in doing so, I respond by reflecting on the issues raised in the chapters and comment on the contributions to the field as I connect to my own work. I will attempt to reflect on them with an eye on reflexively considering the contribution of this collection as a whole. I use the term reflexively intentionally, as with it I mean a particular practice of reflecting upon something with the explicit purpose of looking, and moving, forwards (Siry & Martin 2014). Thus I focus on what can be learned from these chapters in order to reframe, reimagine, and ideally transform science education research and practices.

C. Siry (✉)
University of Luxembourg, Esch-sur-Alzette, Luxembourg
e-mail: Christina.siry@uni.lu

K. Otrel-Cass et al. (eds.), *Cultural, Social, and Political Perspectives in Science Education*, Cultural Studies of Science Education 15,
DOI 10.1007/978-3-319-61191-4_15

As a cultural studies researcher grounded in critical perspectives, I situate science education as cultural enactment and thus something that 'should be analyzed in relation to other cultural dynamics and social and historical structures' (Kincheloe and McLaren 2002, p. 111). The projects my team and I engage in share a focus on examining the dynamics and structures at play in the production of science education, especially at the preschool and primary school levels. In doing so, we seek to reveal and highlight the resources that children bring to their science education experiences that mediate the meanings they make and their embodied engagement with science education, much of which is non-verbal. Thus we work towards developing and utilizing 'resource-rich approaches' to science education (e.g. Siry, 2011) as we situate ourselves and our work to embrace a variety of theories and multimodal methodologies to shine lights on the often unexpected, complex, ways in which children demonstrate their understandings. In doing so, we gain new understandings on the embodied, collective, and profound ways in which children's resources serve to mediate their experiences in science education. It is with this particular research agenda in mind that I read the individual chapters of this book and interpret the book as a whole, and thus my comments that follow consider how 'looking through the windows of the actual' (Greene 2009, p. 138) contributes towards imagining, and hopefully realizing, alternative realities in science education.

The editors of this book state that part of their purpose in writing this book was to "trouble existing frameworks and conditions for researchers, teachers and learners in science classrooms" (p. 2), and thus I take a particular view on this volume building upon the challenge of 'troubling'. My comments in this piece begin with a reflection on the value of such diverse perspectives on science education as are presented in this book. Then I have organized my response in this afterword around three main points that I take away from this edited book as a whole which I have grounded in questions: First, what does the act of troubling give to the field? Second, how can cycles of production and reproduction be disrupted to move towards transformation? Third, what are the implications for locally contextualized research practices with a global reach? Perhaps my reflections on these questions in this afterword provide points for you the reader to also consider your own work with the new lenses offered by these science education scholars from Nordic countries, as I have.

15.1 Diverse Perspectives on Science Education

> Imagination, after all, allows people to think of things as if they could be otherwise; it is the capacity that allows a looking through the windows of the actual towards alternative realities. (Greene 2009, p. 138)

The field of science education will no doubt benefit from a book such as this one that holds as a central purpose the *troubling* of the field with an explicit focus on shining different theoretical lights on various aspects at play in the practice of science teaching, learning, and research. Kenneth Tobin (2008) has elaborated the metaphor that theories can be seen as lights that we shine on a given moment, with a theory highlighting some parts of a particular context but, as with shining a light, obscuring other parts at the same time. Thus he advocates for bricolage approaches to research drawing on diverse theoretical frameworks, and it is precisely the value of such different 'lights' that is underscored through the chapters in this book. As each of the perspectives herein highlights distinctive aspects of the possibilities for critical analyses and consideration, the value of drawing on diverse theoretical frameworks for highlighting different aspects of science education research becomes even more salient.

Science, as a discipline, is embodied; it is something that is done. The artificial separation of the mind and body, as is often produced in science education policy, practice, and research, needs to be brought into focus to move towards a reconceived notion of science education grounded upon the interplay between the senses and the material world, between the body and the mind. This book emphasizes the need to go beyond the historical constructions of science as neutral and thus push back on prevailing perspectives of science as merely based on information that is processed only in the mind. In their contribution to this book, Lotta Leden and Jonna Wiblom emphasize this prevailing idea that 'school science culture, as it is commonly described, has its foundations in ready-made facts and "the scientific method". It seeks to provide insights into a canon of essential scientific phenomena and enculture students through pre-professional science training'. A first step in pushing back at this prevalent notion is to 'reveal or unpack the (antagonistic) values imbedded in science teaching' (p. 70) as Helen Hasslöf and Iann Lundgaard suggest in their chapter and thus move towards an understanding of the complexities and embodied nature of science teaching and learning.

The unfolding of science education practices in classrooms is historically, socially, culturally, and politically constructed, and it is important to consider these forces in the troubling of the field. One example of this is evident when Lars Bang situates notions of science inquiry in his contribution to this volume through a critical analysis of John Dewey's scientific inquiry and its legacy. In doing so, he looks back to the past to move forwards to the future, as he argues for reconceptualizing the notion of 'inquiry' as a flow. Words are not stable constructions, and thus it is critical to surround the words that we use with meaning, and, in science education, terms such as 'hands-on', 'inquiry', and even the ubiquitous 'experiments' need to be unpacked with specific elaborations to arrive at an understanding of what is meant by them in actuality in order to reconstruct something different.

Leden and Wiblom also underscore this dilemma of how science education unfolds in classrooms, as they emphasize '...the sharp distinction between producing and reproducing facts plays out as a mere chimera of the everyday puzzle-solving activity of "real" scientists' (p. 13). Engaging in troubling allows for critical reflection on the goals of science teaching policies and practices in order to move forwards towards transformation, a transformation that hopefully recognizes the immense resources that children bring to bear on their science education and that aligns more closely with building on these resources as one of the goals of science education. In doing so, however, we must take care to not fall into the trap of uncritically trying to solve the 'science education problem' too quickly, as Maria Andrée, Lena Hansson, and Malin Ideland so clearly discuss in their chapter, as this can open up the potential for 'various economic and ideological interests to enter the classroom without being critically scrutinized' (p. 76).

The need for transformation is a pressing one, as there is a misalignment between the practice of science education as it tends to play out in classrooms and the actual practices of children as well as of scientists themselves. Yet there is a boundary that is created by the authority of the subject matter. Gerd Johansen, Gudrún Jónsdóttir, and Stein Dankert Kolstø examine interactions around a subject matter in electrochemistry in their chapter that is '…very well established – it "works" and thus has authority in itself' (p. 125). Troubling can lead to a questioning of the previously unquestioned 'authorities' such as the subject matter, the curriculum, the textbooks used, and the contributions of these authors emphasize that traditional perceptions and authorities can be questioned, and, I would add, *should* be questioned. With this in mind, I now turn to look more precisely at what can be accomplished through the act of troubling, paying particular attention to further elaborating the contributions of this volume on Nordic research in science education.

15.2 On Troubling and Being Troubled: What Does 'Troubling' Give to Our Field?

> We also have our social imagination: the capacity to invent visions of what should be and what might be in our deficient society, on the streets where we live, on our schools. (Greene 2000, p. 5)

My use of Greene in this chapter emerged from the overall inspiration I received from this book to envision something *different.* Different through a diversity of methodologies and theories that ground a reconsideration of that which is present in current pedagogical and research practices in science education. Engaging in the act of troubling implies taking a particular questioning stance in research, a "theoretical, pedagogical, and political stance that to question how thought is thought, or how remembering is remembered, is in and of itself an important practice for attending to, rather than following a complicity with, (feminist) regimes of truth" (Braithwaite et al. 2004, p. 13). This quote about troubling speaks specifically to troubling 'regimes of truth' that are grounded in feminist perspectives, as I draw inspiration from those who have troubled social constructions in educational research, including gender, sexuality, and identity (e.g. Kumashiro 2001) and use these to connect to the social constructions being troubled in this volume as relevant to science education. As the authors focus on particular social, political, and material perspectives, the act of troubling serves to unsettle and to deconstruct taken-for-granted assumptions in our field. The assumptions troubled through the chapters of this book include the sanctity of the science lab, the focus on individualized successes, and the ill-conceived prevalent notion of science as neutral, among others. This act of troubling science education pedagogy and practice leads, in the different chapters, to an emphasis on the inherent problems that result from an artificial segregation of thinking, doing, and feeling in curricula, teaching practices, and research. In this troubling there is a focus on reconceptualizing engagement in science, as, for example, is stressed in Kathrin Otrel-Cass's last chapter which seeks to "reconceptualise who we are and how we make sense of the world and the recognition that the body-mind is present, imbalanced and complex" (p. 193). It is precisely this reconceptualizing with a focus on the direct implications that weaves through the book, as the authors trouble assumptions relative to the goals of science education. Marianne Løken and Margareta Serder reflect on the contribution of the chapters in their 'in-between chapter' leading into the third section, and they highlight how the authors seek to disrupt reductionist analysis and a positivistic scientific stance that appears to hold the goal of science 'as rational, logical, disembodied, objective, and value-free knowledge. They also have in common a more or less explicit critique of dichotomies, whether it is body / mind, nature / culture, or human / nonhuman' (p. xx).

Troubling is essential to theorizing (Braithwaite et al. 2004). It is a tool to problematize and ideally reconstruct and reimagine something that is different. The authors of these chapters use troubling to push back on notions of science education and science education research, as objective, rational and logical, as they dismantle binaries and dichotomies. My work is grounded in critical theoretical perspectives, and a central aspect of my research is considering how science education is taught and what the understandings are that emerge in the teaching and learning of science, specifically at the primary school level. One of the contributions that troubling offers is to unravel the discourses, representations, and ideological perspectives in the field, but it is important to recognize that such unravelling becomes complicated by the 'taken-for-grantedness of the meanings promoted in these representations and the typically undetected ways these meanings are circulated into everyday life' (Kincheloe and McLaren 2002, p. 101). The authors in this book are well placed to go beyond the theories and tools readily available, as they draw on a variety of genres and approaches to examine the micro-, meso-, and macro-levels of social life through a lens on science education.

Research is a human practice; as researchers we interact with and interpret the world. Troubling in this process of interpretation can support questioning what it means to 'teach in ways that challenge the different forms of oppression…what it means to address our resistances to discomforting knowledge, and what it means to put uncertainties and crisis at the center of the learning process' (Kumashiro 2002, cited in Dennis 2017, p. 8–9). Marie Ståhl uses troubling to highlight different visions of science teaching and learning in her chapter, and indeed, one of the contributions of troubling can be to support teachers and researchers in reflecting on preconceived notions of what it means to teach and what it means to learn. School science practices are a 'culture with patterns that are maintained and recreated, patterns often invisible through its regular everydayness' as Anna Jobér clearly explains in her chapter. The production of such culture 'generates knowledge, shapes values, and constructs identity' (Kincheloe and McLaren 2002, p. 95), but there is a danger in the production, and reproduction, of hegemonic ways of perceiving of a field such as science education. Troubling supports making visible these patterns and related taken-for-granted assumptions associated with the cultural production of science teaching and learning. Such cultural production is connected with reproduction, and this production/reproduction occurs in a cyclic process. But often this process is an oppressive one, far removed from students' everyday worlds, perspectives, and understandings. And so I ask; how can we disrupt this cycle? What are the possibilities for transformation?

15.3 Theories and Contingencies: How Can We Move Towards Transformation?

> There can be no final solution; but there is time – always time – to reject somnolence, to choose to begin. (Greene 2010, ¶ 3)

Theories are a guide to the social sphere (Kincheloe and McLaren 2002). Thus, a theory is not a determinant but rather a tool to help explore the world. In this exploration there can be an awakening, a choice to begin to transform. Theories evolve and change as they are engaged with. Several chapters in the book draw on Karen Barad's work (e.g. 2007), and these theoretical perspectives underscore that the world is 'an entangled assembly of things, where the identity, constitution, and even well-being of one entity is contingent on others' (Bazzul 2016, p. 64). Experiences in the world are an entanglement of embodied and emotional experiences – sensing and making sense and engaging with others and with materials – these are central aspects of the processes of science education. As these become unpacked through the use of different theoretical perspectives, it is my hope that there is not an arrival at a final solution to the 'problem of science education' that Andrée, Hansson, and Ideland write about but rather, that, as Greene's quote above emphasizes, there is a choice to begin, to begin to deconstruct through the act of troubling, always, ideally, with a goal of reconstructing something that is more just, more equitable, and more attuned to the resources that children draw on in their inter- and intra-actions.

In Chap. 2, Kathrin Otrel-Cass emphasizes that the researchers involved in this volume decided to create a book to critically review cultural, social, and political perspectives on science education'. These different perspectives on one hand stand alone as individual chapters in this volume, while at the same time they come together and serve to question – to unsettle – the taken-for-granted assumptions, for example, the science lab as something that is so common it is often not brought into question (e.g. Johansen's chapter). As different assumptions are brought to light, they can be unpacked, and as a critically grounded researcher, my hope is that in doing so there can be a transformation that reconstructs something different and something new. Science education is a complex, holistic practice, and through questioning, highlighting, and troubling perhaps, it is that there can be a consistent push back against individualized neoliberal foci that expect a 'best practice' for a value-free science education.

The chapters in this book bring a richness of ideas and voices with a diversity of perspectives on the nature of engaging in the complex practice that is science education. Clearly it is impossible to represent the full complexity of people's participation in the practices of science education, but by shining different theoretical lights on the issues in the field, new perspectives are gained. Diverse theoretical frameworks mediate the emergence and evolution of different perspectives; this is brought to the fore in Auli Arvola Orlander and Ståhl's chapter as they consider their readings of data excerpts and write "we can draw a conclusion that our diffractive reading has contributed to the understanding of our excerpt in a different way" (p. 150). Transformation can occur by first troubling ideologies and discourses in order to uncover, and ideally dismantle, the ideologies that artificially define what *is* and *is not* possible in education and rather welcome differences to learn from them and move forwards towards a plurality of perspectives and approaches. Orlander and Ståhl do just that as they ask "what connections, interferences and sensations emerge" in elaborating their diffractive readings of narratives in the field (p. 145). Such a question facilitates critical reflection and analysis and can situate researchers to work towards new understandings and approaches within science education.

15.4 Locally Contextualized, Yet Globally Relevant: How Can the Two Be Bridged?

> There is a sense in which imagination and desire can feed the recognition of the need to transform and, perhaps, the passion to change… Imagination alters the vision of the way things are; it opens spaces in experience where projects can be devised, the kinds of projects that may bring things closer to what ought to be. (Greene 2009, p. 141)

As mentioned in the introduction to this book, I am one of the coeditors of the journal *Cultural Studies of Science Education* and the related book series in which this volume has been published. When I first met with the editors and discussed their vision for this book, I was intrigued by the focus on research in Nordic countries and the diversity of representations the editors sought to give a place to in such a book. Further intriguing was the use of troubling as a practice to push back on assumptions. Drawing on those who have troubled issues of heteronormativity in other research fields, the act of troubling seeks to disturb the illusion that curriculum is '…value neutral, impartial, and therefore above reproach' as it 'offers the opportunity to look locally and contextually, using global and local theories to make sense of and where necessary, take action' (Francis 2017, p. 8). Those reading this book might begin their reading by wondering what the relevance can be of research particular to Nordic countries as applicable to practices in other contexts. What is the value of troubling science education in Norway, Denmark, and Sweden to classroom practices in Chicago? Or Paris? Or Singapore? Or Capetown? I suggest that it lies precisely in the process of troubling, the beginning of questioning, and the resistance of somnolence that Greene writes about introduced earlier. While the issues faced by educators in Norway may be the same, or perhaps be very different, than those in other contexts, the process of troubling, of questioning to see better what is present and reflexively considering what is possible, is fundamental to moving towards transformation.

Altering 'the vision of the way things are' as Greene presents can begin by taking a critical view on the constructive qualities of time, space, and place in science education. Orlander and Ståhl emphasize that "one cannot, as is usual in research, take a step back and reflect on what one sees and thus be outside the phenomena" (p. 142). It is precisely this point of being 'inside' a phenomenon that can mediate reflection and recognition of 'the need to transform' that Greene underscores. Once this need is present, troubling further supports considering how specifically to conceptualize something that is new and different. Jobér emphasizes that "everything is connected in all possible directions, though time and space in webs and nodes" (p. 20). This is further underscored when there is an emphasis in research on experience and subjectivity, which can support taking a view on politics and perspectives of difference. These perspectives can emerge from taking a relational and contextual view to uncover and learn from contextualized components of our work with others built upon connectedness and relations. Thus, it is critical to consider the implications of this work to other contexts in order to situate lessons learned as locally and contextually relevant. Involving the participants of research, as Peer Daugbjerg and Martin Sillasen's chapter explores, can emphasize "...the relevance of designing PDPs so that the participants' expectations are brought forward and included in the design process" (p. 110). Indeed, in opening ourselves up to truly listen to our participants, we can move towards gaining different perspectives on experiences in social life. Ståhl stresses that "if we want students to be interested in science we must listen to them and highlight their interests and commitments that obviously exist" (p. 175).

One of the aspects of this volume as a whole that is critically important are the overlaps of practices and approaches that are present in the different chapters. The overlaps between citizenship and practical work in school science that Johansen, Jónsdóttir, and Kolstø discuss and the embeddedness of connections between the social, cultural, and material aspects of school science, as Løken and Serder elaborate, for example, underscore the value of considering connections and relations as central aspects of science education. Thus, this volume, while locally contextualized in Nordic countries, urges us to open spaces in time and spaces in place that allow for considering connections and relations and to imagine the new and the different in science education in our own local contexts.

15.5 In Closing

> Imagination, intention: Neither is sufficient. There must be a transmutation of good will, of what I call wide awakeness into action. (Greene 2010)

The authors of this book have brought together diverse perspectives to trouble science education, and as Orlander and Ståhl described in their chapter, there is a particular view on "how entwined and inseparable the cultural, social, political, and material perspectives are" (p. 151). It is precisely this entwined entangledness that is brought to light in this book, as the authors refuse "to take for granted the distinction between the social and cultural, and the human and non-human" as Løken and Serder write about (p. 136). In refusing this taken-for-grantedness, there can be new lights shined on research, with the goal of reaching beyond that which is known, to move towards action. I began this afterword with a quote from Greene that elucidates the value of imagination to create alternative realities. An equitable, just, and accessible science is the alternative reality that I hope for. 'To be human is to have ongoing relations with everything non-human…'. Imagination can support creating, or rather, cocreating science education practices and processes that recognize science education as something different – different in that it draws together the mind and the body, the individual and the collective, and the practices and resources that children bring. In rejecting somnolence, as Greene calls for, there is a wide-awakeness nurtured, one that can mediate beginning, and beginning again, which is in my opinion precisely the central contributions of the chapters of this book. The authors herein have presented inspiring chapters for reflection and consideration that can propel us forwards into such wide-awakeness, as we continually deconstruct and challenge our assumptions, to be ever vigilant of *what is*, in order to bring to question and trouble, to imagine the possibilities of *what can be*, and to create spaces for creating something new.

References

Barad, K. (2007). *Meeting the universe halfway. Quantum physics and the entanglement of matter and meaning*. Durham/London: Duke University Press.

Bazzul, J. (2016). *Ethics and science education: How subjectivity matters*. Dordrecht: Springer.

Braithwaite, A., Heald, S., Luhmann, S., & Rosenberg, S. (2004). *Troubling Women's studies: Pasts, presents, and possibilities*. Ontario: Sumach Press.

Francis, D. (2017). *Troubling the teaching and learning of gender and sexuality: Diversity in South African education*. Dordrecht: Springer.

Greene, M. (2010). Prologue to art, social imagination and action. *Journal of Educational Controversy, 5*(1.) Article 2.

Greene, M. (2009). Teaching as possibility: A light in dark times. In S. L. Macrine (Ed.), *Critical pedagogy in uncertain times: Hope and possibility* (pp. 137–150). New York: Palgrave Macmillan.

Greene, M. (2000). *Releasing the imagination: Essays on education, the arts, and social change*. San Francisco: Jossey Bass.

Kincheloe, J., & McLaren, P. (2002). Rethinking critical theory and qualitative research. In Y. Zou & E. T. Trueba (Eds.), *Ethnography and schools: Qualitative approaches to the study of education* (pp. 87–138). Maryland: Rowman & Littlefield Publishers.

Kumashiro, K. (2001). "Posts" perspectives on anti-oppressive education in social studies, English, mathematics and science. *Educational Researcher, 30*(3), 3–12.

Kumashiro, K. (2002). Troubling education: Queer activism and anti-oppressive pedagogy. New York: Routledge Falmer.

Siry, C., & Martin, S. (2014). Facilitating reflexivity in preservice science teacher education using video analysis and cogenerative dialogue in field-based methods courses. *Eurasia Journal of Mathematics, Science, and Technology Education, 10*(5), 481–508.

Siry, C. (2011). Exploring the significance of resource-rich views in science education. *Cultural Studies of Science Education, 6*(4), 1019–1029.

Tobin, K. (2008). In search of new lights: Getting the most from competing perspectives. *Cultural Studies of Science Education, 2*(3), 227–230.

Index

K. Otrel-Cass et al. (eds.), *Cultural, Social, and Political Perspectives in Science Education*, Cultural Studies of Science Education 15,
DOI 10.1007/978-3-319-61191-4

MIX
Papier aus verantwortungsvollen Quellen
Paper from responsible sources
FSC® C105338

If you have any concerns about our products, you can contact us on
ProductSafety@springernature.com

In case Publisher is established outside the EU, the EU authorized representative is:
Springer Nature Customer Service Center GmbH
Europaplatz 3, 69115 Heidelberg, Germany

Printed by Libri Plureos GmbH
in Hamburg, Germany